SOLAR-POWERED ELECTRICITY

SOLAR-POWERED ELECTRICITY

A Survey of Photovoltaic Power in Developing Countries

Bernard McNelis, Anthony Derrick and Michael Starr

Intermediate Technology Publications
in association with UNESCO

Published by ITDG Publishing
The Schumacher Centre for Technology and Development
Bourton Hall, Bourton-on-Dunsmore, Rugby, Warwickshire CV23 9QZ, UK
www.itdgpublishing.org.uk

First published in 1988
Reprinted 1992
Print on demand since 2004

ISBN 0 946688 39 7

A catalogue record for this book is available from the British Library

ITDG Publishing is the publishing arm of the Intermediate Technology Development Group.
Our mission is to build the skills and capacity of people in developing countries through the dissemination of information in all forms, enabling them to improve the quality of their lives and that of future generations.

IT Power Ltd, The Manor House, Chineham Court, Lutyens Close, Chineham, Hampshire RG24 8AG, UK

Printed in Great Britain by Lightning Source, Milton Keynes

CONTENTS

1.	**INTRODUCTION**	**1**
	1.1 Background	1
	1.2 Objectives and scope	1
	1.3 Approach	2
2.	**OVERVIEW OF PHOTOVOLTAIC TECHNOLOGY AND APPLICATIONS**	**5**
	2.1 Why photovoltaics for developing countries?	5
	2.2 Photovoltaic technology	5
	2.3 Systems and applications	11
	2.4 Photovoltaic manufacturers, markets and prospects	12
3.	**WATER PUMPING**	**19**
	3.1 Introduction	19
	3.2 Principal case study — Mali	22
	3.3 Other water pumping projects	27
	3.4 Conclusions	28
4.	**PHOTOVOLTAIC REFRIGERATORS FOR RURAL HEALTH CARE**	**33**
	4.1 Introduction	33
	4.2 Principal case study — India	38
	4.3 Other PV refrigeration projects	39
	4.4 Conclusions	45
5.	**LIGHTING**	**51**
	5.1 Introduction	51
	5.2 Principal case study — South Pacific	52
	5.3 Other lighting projects	54
	5.4 Conclusions	55
6.	**RURAL ELECTRIFICATION**	**59**
	6.1 Introduction	59
	6.2 Principal case study — Indonesia	62
	6.3 Other rural electrification projects	65
	6.4 Conclusions	66
7.	**OTHER APPLICATIONS**	**71**
	7.1 Agricultural applications	71
	7.2 Water treatment	72
	7.3 Telecommunications	72
	7.4 Cathodic protection	74
	7.5 Unusual applications	74

8. **CONCLUSIONS** **77**
8.1 Summary of experience 77
8.2 The technology 81
8.3 The economics 82
8.4 Social and institutional factors 82

9. **RECOMMENDATIONS** **85**
9.1 Identification of appropriate applications 85
9.2 Strategic approach to development 85
9.3 Staged development 86

1. INTRODUCTION

1.1 Background

The provision of adequate supplies of energy in suitable forms and at acceptable prices is an essential prerequisite for most development activities. Energy supply problems were brought to the forefront of attention in 1973 with the steep increases in the price of oil. The impact of higher oil prices has fallen particularly heavily on developing countries, as the cost of their energy imports constitutes a much higher proportion of export earnings than for industrialized countries.

One of the ways the industrialized countries responded to the energy crises was to initiate, or greatly expand, research, development and demonstration (RD and D) programmes in new and renewable sources of energy and in energy conservation. Many developing countries also introduced energy RD and D programmes, but continuing financial constraints have inevitably limited the scale of these activities. There have nevertheless been many projects and development programmes initiated in developing countries over the last 10 years or so, many with technical and financial assistance from the industrialized countries.

One important area of renewable energy RD and D has been in solar photovoltaics, the direct conversation of solar energy into direct current electricity by means of solar cells. Photovoltaic (PV) systems first came into prominence in the late 1950s for powering space satellites. After the oil crisis in the early 1970s, PV systems were developed for a wide range of terrestrial applications. When correctly designed and installed, PV systems will operate for many years, requiring little supervision and only occasional simple maintenance. They need no fuel supplies and give rise to no pollution.

At first the costs of PV systems were very high, but with improved technology, cheaper materials and higher volume production, prices have been steadily falling in real terms. The stage has now been reached when PV systems are both technically and economically suitable for many applications, particularly those involving relatively small amounts of power in remote locations, where the cost of operating and maintaining a conventionally powered system is high.

Over the last 10 years, PV systems have been installed in developing countries to supply power for water pumping, refrigeration, lighting, village electrification, communications and other applications. Many of these systems were installed as part of development and demonstration projects and it is now appropriate to make a comprehensive evaluation of the experience gained.

1.2 Objectives and scope

The main purpose of this survey is to review the present state of knowledge regarding photovoltaic applications in developing countries and to assess future prospects. Many lessons have been learned from projects including those where photovoltaic-powered systems did not perform as well as expected. It is now vital to disseminate the information available so that valuable resources are not wasted.

In addition to reviewing the experience gained with PV systems in developing countries, the report also provides advice on the selection of appropriate equipment, taking into account the various technical, economic, social and institutional factors involved. The recommendations are intended to help decision makers identify which photovoltaic applications are suitable for the specific conditions obtaining in the regions for which they are responsible and to give guidance on how to implement the necessary projects.

A further objective of the report is to identify areas where further development and demonstration activities are needed. Such activities include the training of local personnel in the design, installation and evaluation of systems and the actions needed to provide the basis for local manufacturing facilities, as well as field demonstrations for selected applications.

Chapter 2 first addresses the question of why photovoltaics are of particular interest for developing countries and then presents a summary of photovoltaic technology and applications. Chapters 3 to 6 deal with each main applications area in turn, starting with water pumping and proceeding to vaccine storage and other medical applications, lighting and rural electrification. Other applications are considered in Chapter 7.

The conclusions are presented in Chapter 8, with an overall summary of the experience followed by specific conclusions relating to the technology and the economics. The applications that are considered to be the most appropriate for developing countries are then identified. The chapter concludes with a review of the issues which are important for the successful implementation of a photovoltaic project.

The recommendations arising out of the study are listed in Chapter 9. The first section covers the methodology that should be adopted for identifying appropriate PV applications. The second section deals with the overall approach to the development of photovoltaics in a developing country, and the final section lists the priority topics for research, development and demonstration needed for the implementation of photovoltaics in developing countries.

1.3 Approach

The information on which this survey is based comes from two main sources: published reports, technical papers and articles; and in-house knowledge available with the staff of IT Power, built up over many years of experience of photovoltaic projects worldwide.

In considering each application, first a general introduction is given, covering the main issues involved and reviewing the status of development worldwide. This provides the background for the principal case study based on a specific country or region and covering the technical, economic and social/institutional aspects in detail. This is followed by a briefer review of significant projects in other countries.

The conclusions arising from the review of the experience for each application are then listed and discussed in relation to the following headings:

(a) Technical
- Reliability
- Availability
- Durability
- Ease of operation and maintenance

(b) Economic
- Capital cost

Operation and maintenance costs
Life cycle cost/benefit
Comparison with conventional alternatives

(c) Social/institutional
Availability and quality of institutional support
Demand for product or services
Compatibility with social requirements of user
Availability of skills needed for operation and maintenance.

2. OVERVIEW OF PHOTOVOLTAIC TECHNOLOGY AND APPLICATIONS

2.1 Why photovoltaics for developing countries?

Energy is needed for practically all the activities that are basic to human survival, such as cooking, water pumping and food production. After basic needs are satisfied, further energy is required to improve the quality of life, through lighting, transport, telephone communications and consumer tools such as refrigerators, radios and televisions. As a country develops, still further inputs of energy are required for industries and for commercial and public buildings. In urban areas, the necessary energy supplies may be readily provided through oil products, coal and networks for electricity and natural gas. In rural areas, traditional sources of energy, principally firewood, agricultural residues and cattle dung, continue to be of major importance, supplemented by commercial sources such as electricity and oil products in areas where the physical infrastructure makes this possible.

The majority of the population of all developing countries live in the rural areas. The combined effect of population growth and supply problems of commercial fuels is putting ever-increasing pressure on the traditional fuel supplies. Deforestation resulting from over-cutting of trees, sometimes aggravated by long-term climatic changes, is becoming a major problem in many countries. The use of agricultural residues and cattle dung as fuel reduces the amount of nutrients returned to the soil.

Photovoltaic systems are widely recognized as an attractive means to address some of the rural energy problems, since they offer the following advantages:

- Being built up from solar cell modules, they are able to provide relatively small amounts of electrical power at or close to the point of demand
- No fuel requirements
- Relatively simple operation and maintenance requirements, within the capability of unskilled users
- No harmful pollution at the place of use
- Long life with little degradation in performance.

The remainder of this chapter provides a summary of photovoltaic technology and a general review of systems, applications, markets and prospects. This is intended to provide the background for the subsequent chapters of this report which deal with specific applications in developing countries.

2.2 Photovoltaic technology

Brief history

The photovoltaic effect was first observed by the French scientist Becquerel in 1839 who noticed that when light was directed onto one side of a simple battery cell, the generated current could be increased. Work on the photovoltaic properties of selenium in the 1870s led to the first selenium photovoltaic cell in 1883. The photo-sensitive properties of copper and cuprous oxide structures were discovered in 1904. By 1905, it was known that the number and energy level of electrons emitted by a photosensitive substance varied with the intensity and wavelength of the light shining on it.

In the years that followed research work continued with the objective of developing practical photovoltaic devices. Selenium and cuprous oxide photovoltaic cells were developed, leading to several applications including photographic exposure meters and other small light sensors. By 1941, selenium devices had been developed with a light-to-electricity efficiency of about 1 per cent. A new technique was later developed, known as a 'grown p-n junction', which enabled the production of single-crystal cells. Using doped silicon crystals, American research workers in the mid-1950s were able to achieve solar conversion efficiencies up to 6 per cent.

Western Electric began to sell commercial licences for silicon photovoltaic technology in 1955 and there were some attempts to develop practical systems for powering specialist equipment in remote areas. It was not until the late 1950s however that crystalline silicon solar cells were developed with high enough conversion efficiencies for their use in power generators. A major impetus for the development of these cells was the space programme. The first solar-powered satellite, Vanguard I, was launched by the USA in 1958. Practically all satellites launched since then have been powered by solar arrays made up of many thousands of crystalline silicon photovoltaic cells.

Following the 1973 oil crises, interest in photovoltaics as a terrestrial source of power increased greatly and many countries, including several developing countries, instituted photovoltaic research, development and demonstration activities as part of wider energy research programmes. Total world expenditure from all sources on photovoltaic research, development and demonstration activities is probably running at between $200 and $300 million per annum.

Over the last ten years, there has been more than a tenfold reduction in the real price of photovotlaic modules. This has been achieved through a combination of improved cell technologies and larger manufacturing volumes. Starting from virtually zero in 1974, sales of photovoltaic systems have grown to about 25 MWp in 1985, with a total value of at least $800 million. Worldwide there are over 20 module manufacturers of significance and there are several times this number of firms designing and marketing photovoltaic systems using bought-in components. Research efforts are continuing on a broad front to develop better photovoltaic devices and lower cost systems.

Much has already been achieved and many thousands of systems are operating reliably today, ranging in size from a few watts to several megawatts. From being an exotic, highly expensive technology for very specialized situations, photovoltaic generators are now an appropriate solution for a growing number of applications and in time could become a major factor in world energy supplies.

Solar radiation

The energy generated continuously by the sun is radiated as a stream of photons of various energy levels. At a point just outside the earth's atmosphere, the intensity of the solar radiation incident on a plane normal to the sun's rays is known as the solar constant. The average value of this is 1353 W/m^2, with seasonal variations due to the elliptical nature of the earth's orbit. As the solar radiation passes through the earth's atmosphere, a considerable amount is lost by scattering and absorption, some wavelengths being affected more than others. The amount of energy lost depends on the path length of the direct solar beam through the atmosphere and the amount of dust and water vapour at the time.

The solar irradiance at ground level is made up of a direct component and a diffuse component. The sum of these two components on a horizontal plane is termed the 'global irradiance'. The diffuse component can vary from about 20 per cent of the global on a clear day, to 100 per cent in heavily overcast conditions. On a clear day in the tropics, with the sun high overhead, the global irradiance can exceed 1000 W/m^2, but in northern Europe it rarely exceeds 850 W/m^2, falling to less than 100 W/m^2 on a cloudy day.

Knowledge of the solar radiation reaching a photovoltaic cell is important, since not only is the total output from the cell dependent on the intensity of the incident radiation, but also different types of cell show varying levels of response to the different wavelengths of incoming energy. To compare solar cells, it is normal to quote the maximum power output in peak watts (Wp) at Standard Test Conditions (STC), defined as an irradiance of 1000 W/m^2, with a reference sunlight spectral energy distribution and a cell temperature of 25°C.

The photovoltaic process

The material most commonly used to make photovoltaic cells for power applications is crystalline silicon, either in mono-crystalline or, more recently, semi-crystalline form. The essential features of this type of cell are shown in Figure 2.1. It is made from a thin wafer of high purity silicon, doped with a minute quantity of boron. Phosphorus is diffused at a high temperature into the active surface of the wafer. The front electrical contact is made by a metallic grid and the back contact usually covers the whole surface. An anti-reflective coating (ARC) is applied to the front surface.

The phosphorus introduced into the silicon gives rise to an excess of what are known as conduction-band electrons and the boron an excess of valence-electron vacancies or 'holes', which act like positive charges. At the junction, conduction electrons from the n (negative) region diffuse into the p (positive) region and combine with holes, thus cancelling their charges. The area around the junction is thus depleted in charge by the disappearance of electrons and holes close by. Layers of charged impurity atoms (phosphorus and boron), positive in the n region and negative in the p region, are formed either side of the junction, thereby setting up a 'reverse' electric field.

When light falls on the active surface, photons with energy exceeding a certain critical level known as the bandgap (1.1 electron-volts in the case of silicon) interact with the valence electrons and elevate them to the conduction band. This process leaves 'holes', so the photons are said to generate 'electron-hole' pairs which are generated throughout the thickness of the wafer in concentrations depending on the intensity and spectral distribution of the light. The electrons move throughout the crystal lattice and the less mobile holes also move by valence-electron substitution from atom to atom. Some recombine, neutralizing their charges, and the energy is converted to heat. Others reach the junction and are separated by the reverse field, the electrons being accelerated to the negative contact and the holes towards the positive. A potential difference, or open-circuit voltage (Voc), is thus established across the cell which is capable of driving a current through an external load.

The current-voltage relationship (I-V) characteristic for a typical cell is dependent on irradiance and temperature, as illustrated in Figures 2.2 (a) and (b). For crystalline silicon cells, when illuminated by light with intensity 1000 W/m^2 AM 1.5 direct spectrum, at 25°C, the open-circuit voltage (Voc) is about 0.6V and the short-circuit current (Isc) about 30 mA/cm^2.

As the cell temperature increases, the current increases slightly as the voltage decreases significantly, in consequence, the maximum power decreases. It is therefore desirable to operate the cells at as low a temperature as possible.

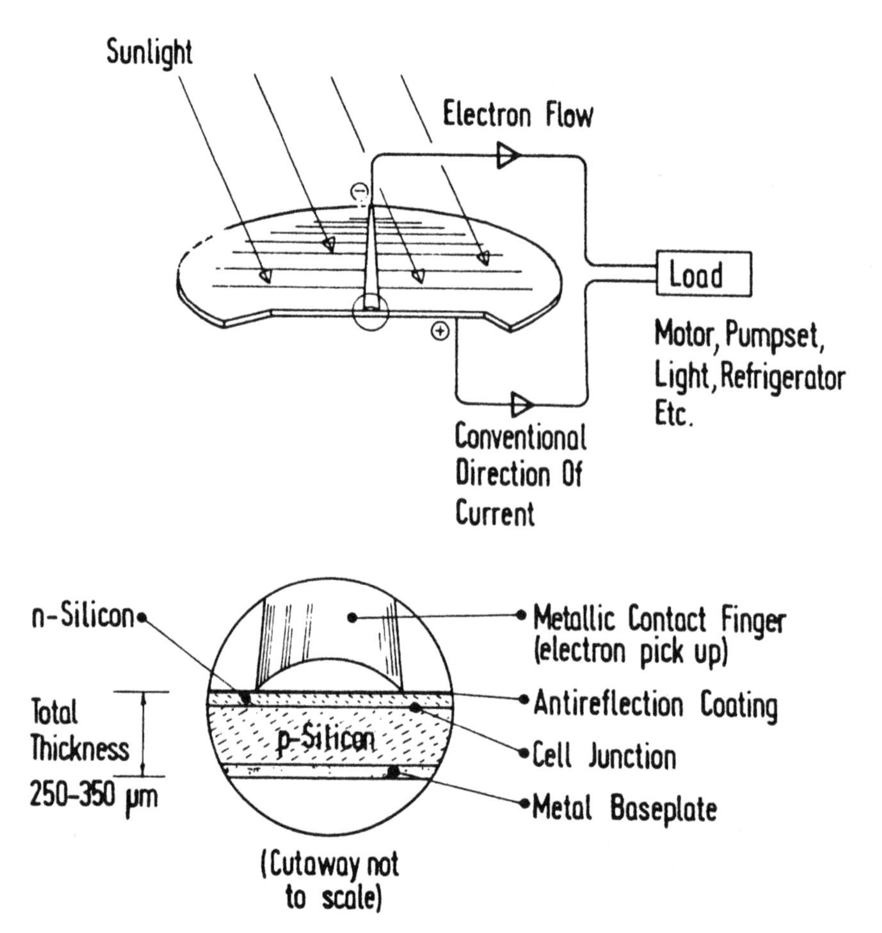

Figure 2.1 Crystalline Silicon Photovoltaic Cell

The cell efficiency is the ratio of the maximum power to the product of gross cell area and irradiance, usually expressed as a percentage. The photovoltaic process, like other energy conversion processes, is subject to a maximum efficiency dependent on the physical characteristics of the materials. The achievement of improved working efficiences, closer to the practicable maximum, is therefore a major objective of research and development work. For example, the maximum practicable conversion efficiency for conventional crystalline silicon cells is about 25 per cent, but the efficiency actually achieved for mono-crystalline cells commercially manufactured is typically about 14 per cent, although 20 per cent has been reported for cells made in a research laboratory.

Crystalline silicon cells

The mono-crystalline silicon solar cell is a highly stable device and is based on well-established semi-conductor technology developed over many years for integrated circuits. Wafers about 250-350 μm thick are cut from long single crystal ingots 75 mm, 100 mm or even 150 mm in diameter. The ingots are sometimes made by the 'float zone' (Fz) process, but more usually the Czochralski (Cz) process is employed, whereby an ingot is slowly drawn out of a melt of doped silicon in an inert atmosphere. The atoms of silicon solidify into a perfect cubic lattice following the structure of a seed crystal. Commercial photovoltaic cells made from the wafers typically have efficiencies in the range 11-15 per cent.

Several groups have developed cast ingot processes which are less energy intensive and which are more tolerant of impurities. A melt of doped silicon is formed in a mould up to 300 mm cube and allowed to solidify under carefully controlled conditions. The resulting ingot has a semi-crystalline structure which is clearly revealed when it is sliced up to form wafers, usually 100 mm square. The resulting solar cells typically have efficiencies in the range 10-12 per cent with some manufacturers even being able to achieve more than 12 per cent using surface passivation or gettering techniques.

All ingot processes, whether for mono or semi-crystalline silicon, have the drawback that they involve sawing to form wafers. This is a time-consuming and wasteful operation, with over half the material lost.

As an alternative to the ingot processes, several research teams have been working for some years on the development of continuous sheet processes, which do not need subsequent sawing. The main problem with all these processes has been to achieve an acceptable quality of crystalline silicon sheet with a sufficiently high rate of production to render the process economic. The only commercial sheet process which has emerged to date is that developed by Mobil Solar Energy Corporation (USA), which involves drawing out a nine-sided thin-walled polygon from a silicon melt. Rectangular wafers are then cut from the walls of the polygon

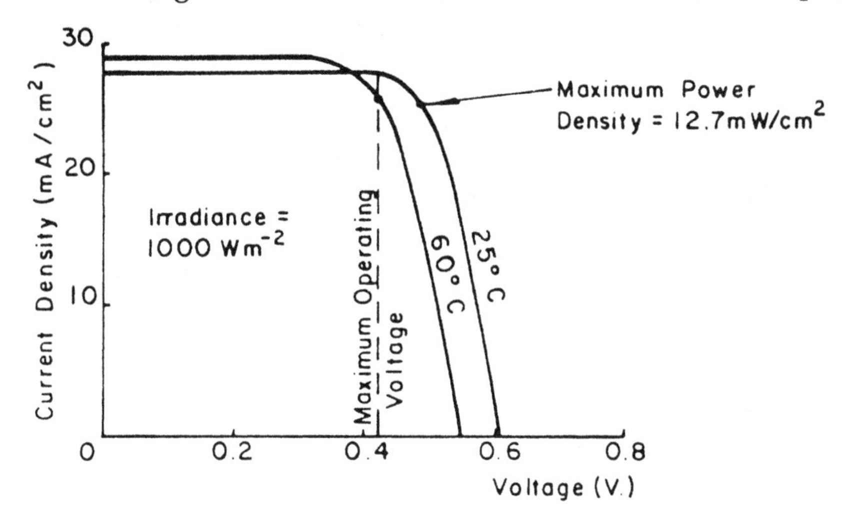

EFFECT OF CELL TEMPERATURE ON V-1 CHARACTERISTIC

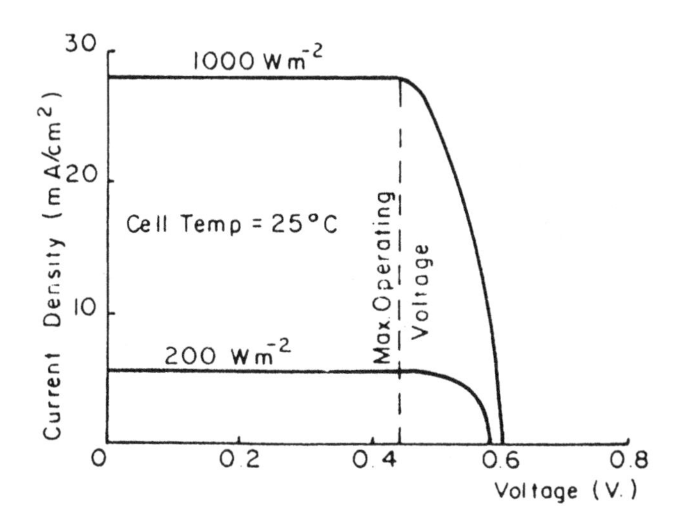

EFFECT OF CHANGE IN IRRADIANCE ON V-I CHARACTERISTIC

Figure 2.2 Typical V-I Characteristic Curves

and made into cells. Cell efficiencies are reported to be comparable with those for ingot processes.

Thin film solar cells

Thin film solar cells — due to low material needs and economic mass production possibilities — have excellent prospects for the future. R and D-efforts are carried out on cells based on amorphous silicon and on polycrystalline materials like IV-VI compounds and chalcopyrites. Amorphous silicon-based thin film devices have entered the consumer electronics market since a couple of years and are at present undergoing commercialization for the power market. Thin film cells based on CdTe/CdS are as well entering the market. Present research efforts are being directed to improving the cell efficiency and — especially with amorphous silicon cells — long-term stability of thin film devices. In early 1986 these devices were beginning to enter the market for developing country applications.

Modules and arrays

Solar cells can be interconnected in series and in parallel to achieve the desired operating voltage and current. The basic building block of a flat-plate solar array is the module in which the interconnected cells are encapsulated behind a transparent window to protect the cells from the weather and mechanical damage. One or more modules are then attached to a supporting structure to form a panel and a number of panels makes up an array field which, together with the balance-of-system (BOS) components, makes up the complete system. The array field may be sub divided electrically into a number of sub-arrays working in parallel. A selection of modules is shown in Figure 2.3. Flat plate arrays are normally fixed, with the modules supported by a structure such that they are orientated due South (in the Northern hemisphere) and inclined at or about the angle of latitude to maximize the amount of solar radiation received on an annual basis. A steeper angle of inclination will enhance the output in winter, at the expense of some reduced output in summer.

For some circumstances, it is appropriate and cost-effective to mount the modules on a support structure that tracks the sun through the day. Given clear sky conditions, the output from the array tracked in this way is more uniform and can exceed that from a fixed array by at least 20 per cent; moreover, the extra output comes in the early morning and late afternoon, the times when demand for grid-supplied electricity is often highest. However, in view of the additional complication and the need for more skilled operation and maintenance, tracking collectors are generally not appropriate for remote sites, where fixed flat-plate arrays are preferable.

Concentrator devices

Although a large number of concentrator photovoltaic devices have been developed, the long-term prospects for this approach are not favourable, at least for high concentration systems for terrestrial applications. Apart from special applications where the concentrator's higher efficiency and potential for providing thermal as well as electrical energy can be exploited to the full, the simplicity and reliability of flat plate modules constitute attractive advantages. It should be noted that in addition to maintaining the tracking system, the optical components of the concentrating system (ie. lenses or reflectors) have to be regularly cleaned, adding to the operational costs. However, until such time as very low cost flat plate systems are developed, low concentration systems (eg. double mirror or Fresnel lens devices) will often be found to offer economic advantages for large installations at places where the necessary skilled maintenance staff are available.

A number of such systems have been built in recent years, including a 6.5 MWp central generating plant built by Arco Solar at Carrisa Plains, California USA, which uses double-mirror concentrating collectors.

2.3 Systems and applications

Market categories

There are currently three main market categories for photovoltaic systems. Firstly, there is the large and growing consumer market, for calculators and other small electronic devices, cooling fans, battery chargers, lights and other small PV systems. Sales in this market are largely dependent on good design, effective marketing and reasonable prices. Secondly, there is the market for professional systems, such as generators for telecommunication links, cathodic protection, navigation lights, military equipment, etc. These systems normally have to be justified on the basis of life-cycle costings using conventional economic criteria, although environmental considerations can often be important. Thirdly, there is the very large potential market for systems which primarily have a social benefit, such as the provision of electricity for remote houses, water supply pumps for villages, emergency telephone links, etc. These systems are generally expensive, but in places where diesel generators or grid extension would be impracticable, the photovoltaic solution can provide important social benefits to the community.

In the longer term, if the very low-cost targets can be achieved, a fourth market category is expected to open up, namely that of grid-connected systems providing electrical power to buildings of all types or serving as central generators.

Figure 2.3 Selection of PV Modules

Stand-alone systems

Most photovoltaic manufacturers now offer a wide range of standard systems, for battery charging, water pumping, street lighting, domestic lighting, refrigeration, electric fencing, alarm and security equipment, remote monitoring, beacons and other navigational aids; the list is constantly growing as other applications are being found. Although some further improvement and demonstration of these systems is continuing, in most cases they can be considered as developed products. Commercial sales are growing steadily to private and public customers, who find that photovoltaic systems provide the most economic or convenient solution to their needs.

For developing countries, the main applications of interest for rural development are:

— water pumping for potable water supplies, livestock watering and irrigation
— vaccine storage in refrigerators and other medical applications
— lighting systems for domestic and commercial uses
— village electrification.

There are a number of other applications which also have their place in rural development, such as radio telephone links, educational television, mechanized milling of rice and other grains, electric cattle fencing and cathodic protection of steel structures and pipelines.

Further information on the technical, economic and social/institutional aspects of the above applications are given in subsequent chapters of this survey.

Grid-connected systems

There has been much discussion of the possibility that photovoltaics will eventually become cheap enough to be economic for grid-connected applications. At present (1988), with oil and coal prices depressed, this seems to be a remote prospect, but in the long term the position could change, particularly for countries rich in solar energy but low in conventional fuels and unwilling (or unable) to introduce nuclear technology. Many countries have made a strong political commitment to encourage the use of renewable energy resources and some have gone further by deciding not to build any new nuclear power plants (eg. Sweden).

Grid-connected systems are simpler and less expensive than stand-alone systems, since they require little or no battery storage. The grid itself can serve as 'storage', with the photovoltaic plant supplying power to or drawing power from the grid depending on the load and solar irradiance. There are a number of grid-connected photovoltaic systems in the USA.

2.4 Photovoltaic manufacturers, markets and prospects

The photovoltaic industry worldwide

Stimulated by the research and development programmes in the USA, Japan and Europe over the last 10 years, a photovoltaic industry for terrestrial applications has emerged in practically every industrialized country and in many developing countries. As may be expected in a new area of technology, there have been a number of failures and disappointments as well as some notable successes. Many of the main photovoltaic manufacturers are subsidiaries of, or are majority owned by, major oil companies, who see photovoltaics as a natural extension of their energy interest, and a field which may become very large in the future. In the USA, there are now some 10 to 15 companies well established in the manufacture and marketing of photovoltaic modules and systems. In Europe, there are about 8 to 10 photovoltaic manufacturers with annual production at least 100 kWp, plus a similar number of smaller organizations more concerned with marketing systems using bought-in components, the so-called OEMs (original equipment manufacturers).

In Japan, at least six companies are active in photovoltaics, including several of the main manufacturers of electrical and electronic goods, such as Kyocera, Mitsubishi, Sanyo, Sharp, Toshiba and Fuji.

A number of developing countries are developing their own photovoltaic industry, notably Brazil, China, India and Pakistan. Locally manufactured photovoltaic modules are not significantly cheaper than similar products made in industrialized countries, but because foreign products are usually subject to import taxes, local manufacturers are protected and are thus able to secure the local market, supplying photovoltaic systems for demonstration projects and professional applications such as telecommunications and water supplies. In this way, local manufacturers are able to build up experience in the design, manufacture, operation and evaluation of photovoltaic systems, in readiness for the day when new technologies can be introduced and become a major energy resource for rural development.

Photovoltaic markets

Total sales of PV systems amounted to about 25 MWp in both 1984 and 1985, indicating that the business is currently worth at least $800 million per annum. The growth in sales over the last 10 years is illustrated in Figure 2.4. The very high growth rate in the 1970s has now settled down to about 30 per cent per annum. The approximate breakdown of the total market in 1984 is set out in Table 2.1.

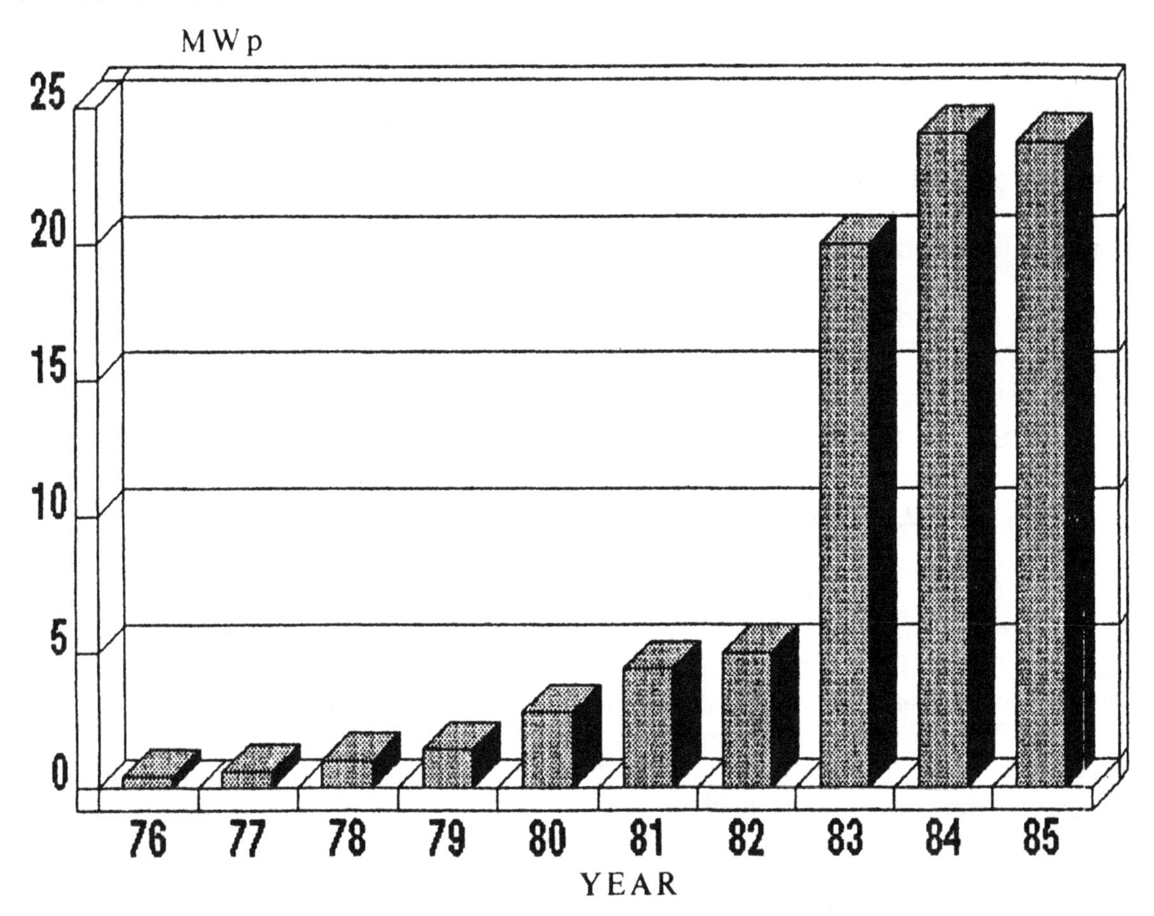

Figure 2.4 Photovoltaic Sales for Terrestrial Applications

Technology and prices

Over the last ten years, the price of photovoltaic modules and systems has been steadily falling in real terms. Module prices for both forms of crystalline silicon are currently around $5 to $7/Wp for large orders (FOB). Bearing in mind that the cells

Application	Sales in MWp
Telecommunications	5.0
Cathodic protection	2.0
PV/Diesel hybrid generators	1.0
Water pumps	0.4
Navigational aids	0.4
Off-grid residential — USA	2.0
Grid-connected residential — USA	0.1
Rural electrification (excluding USA)	0.2
Refrigerators	0.2
Military applications	0.1
Central generators — USA	6.3
Consumer products — crystalline Si	0.1
Consumer products — amorphous Si	7.0
Miscellaneous	0.2
TOTAL SALES IN 1984	25.0

Table 2.1 Estimated Total Sales of Photovoltaics in 1984

account for about 60 per cent of the module price, some further price reductions, possibly down to about $2-$3/Wp, are foreseen through the introduction of cheaper silicon and larger, fully automated manufacturing plants. Much lower costs, even down to $1/Wp or less, are potentially attainable with thin film cells. In view of the large efforts being made world-wide to develop different thin film technologies, it is probable that large-area thin-film cells will become available by 1990 with much improved efficiency and stability compared with current products. Some researchers maintain that crystalline silicon cells could continue to be competitive with thin film processes.

Market prospects are largely dependent on prices of photovoltaics in relation to alternative energy sources, but other factors are important, such as government incentives, availability of finance and the general perception of the technology held by potential customers. Although it is not possible to predict with precision what the future market will be, Table 2.2 indicates what the future sales of photovoltaic systems worldwide might be for two scenarios. The low price scenario assumes that large area thin film cells with adequate performance for power applications start becoming available within the next two to three years; the high price scenario is based on the assumption that the technical targets for thin film cells remain elusive, leaving crystalline silicon as the dominant technology for power applications. There will continue to be a growing market for consumer goods powered by small area amorphous silicon cells, even if prices remain at today's levels.

For the low scenario, with thin film module prices falling to around $1.5/Wp, total annual sales are projected to grow rapidly, from the current level of about 25 MWp to as high as 5000 MWp by AD2000, with continued expansion thereafter. Most of the output would be in and for developing countries for rural electrification and irrigation pumping, using stand-alone systems, but there would also be many applications in industrialized countries for consumer systems, professional systems and remote houses and villages. Grid-connected applications could begin to become a significant market in some countries by the late 1990s.

For the high scenario, with crystalline silicon module prices falling to about $3/Wp and thin film cells not able to compete for power applications, the total market would grow much more slowly, possibly levelling out at about 200 MWp per annum by AD2000. Most of the sales would be for consumer systems and professional systems, with relatively little going to rural electrification, because of the high capital costs involved. However in some countries, there would be good markets among more wealthy private customers for powering isolated houses and for consumer systems, particularly for the tourist and leisure markets. Systems installed by national governments and public utilities would be mainly for applications with high social value. Although the market would be relatively limited, probably only a few megawatt per annum, the benefits to isolated communities would be high.

Year	LOW PRICE SCENARIO			HIGH PRICE SCENARIO		
	Modules $/Wp	Systems $/Wp	Sales MWp/yr	Modules $/Wp	Systems $/Wp	Sales MWp/yr
1984	6.5	12-22	25	6.5	12-22	25
1990	3.0	6-10	100	4.0	8-15	50
1995	2.0	3-7	700	3.0	6-10	100
2000	1.5	2.5-5	5000	3.0	5-9	200

Table 2.2 Projection of PV Prices and Sales 1983-2000 (1985 US$)

Market development

Developing countries have always been considered as a very large potential market but, due to financing problems, actual commercial sales in these countries are at present very small. In fact, the greater part of the systems installed to date in developing countries has been assisted by foreign governments and/or the international aid agencies. Developing countries are rightly concerned to ensure that at the right time photovoltaic technology is transferred to them, rather than find themselves dependent yet again on an imported energy technology. In due course, it is likely that most developing countries will have their own PV industry, but this will take many years to establish, during which time there will be a need to import systems for demonstration projects, professional applications and key community applications.

A significant market in the short term may well be for solar refrigerators and other consumer products for the wealthier sections of society, but in the medium and long terms much larger markets can be expected to develop for professional systems, particulary for telecommunications, village water supplies and generators for police posts and health centres. If system costs can be brought down to about a third of current levels which may well be possible within five to ten years, rural electrification using photovoltaics will become a viable option in many situations, with market potential reckoned in many hundreds of megawatts per annum.

Institutional support

Established institutions already exist in the industrialized countries for co-ordinating and funding energy research, development and demonstration activities, and for disseminating information on the performance, costs and benefits of new technologies. In view of the importance of photovoltaics for applications now and in the future, it is clear that continued institutional support will be needed for this technology. The photovoltaic industry in each country has to compete in world markets and for this there needs to be a sustained programme of technical support, not only for research and development but also for the formulation of appropriate norms and standards and dissemination of information. The recent instability in the price of oil underlines the need for consistent long-term support for new and renewable energy sources such as solar photovoltaics.

Until such time as photovoltaic systems become cheap enough to be economically viable for grid-connected applications, the markets in industrialized countries will remain relatively limited. Apart from the cost factor, there are no major technical or institutional barriers to be overcome, although there is still a need for information on photovoltaics to be disseminated to potential customers and to government officials, many of whom remain unaware of the opportunities offered by this relatively new and formerly rather exotic technology. Some electricity utilities have adopted a rather negative attitude to photovoltaics in the past; they now need to recognize that photovoltaic systems can complement more traditional systems and, in particular, can help solve electricity supply problems in remote areas.

In the developing world, the opportunities for PV are much greater than in the developed world but so are the obstacles to be overcome. A major need in many countries is for an effective institutional base that can monitor, plan and regulate developments in a technology as new and as promising as photovoltaics. The experience available within the industrialized countries could do much to assist developing countries build up the necessary institutional base. Private and public interest need to be brought together to ensure that the following aspects are properly covered:

(a) Technical development
— identification of projects that can be properly justified with reference to all relevant factors
— applied research to develop systems appropriate for local needs
— commercial manufacture and assembly of components and systems

— integration of photovoltaics into related fields (eg. building construction and electronics)
— field installations for demonstration purposes, with monitoring and evaluation

(b) Regulatory aspects
— development and implementation of appropriate norms and standards
— regulations for consumer protection and warranties
— independent testing and certification of components and systems
— planning of integrated projects with no disturbing side effects

(c) Incentives and finance
— tax credits, subsidies and grants
— public and private finance for manufacturers and customers
— co-ordination of aid-funded projects

(d) Training and information
— training of professionals in all aspects of photovoltaics
— public information and advisory services
— liaison with other energy supply agencies.

3. WATER PUMPING

3.1 Introduction

Pumping techniques

Hand- and wind-powered pumps have a long history for lifting water in rural areas for water supply and irrigation. Improved designs of hand pumps that are more efficient and durable and easier to maintain are now being widely introduced. There has also been renewed interest in wind pumps in recent years, with the emphasis on lower cost designs suitable for local manufacture. Petrol and diesel pumps are also being used in some areas, particularly for low-lift irrigation.

In rural areas that have been electrified by grid extension, electric pumps are usually a reliable and relatively low cost option. However, for most rural areas, it will be many years before this alternative is available.

Substantial efforts have been made in recent years to develop reliable and cost-effective solar-powered pumping systems. A number of prototype solar thermal systems have been developed, but none so far offer sufficient reliability, ease of operation and maintenance and cost-effectiveness. Photovoltaic systems on the other hand offer a number of attractive features and, after several years of development, are now readily available in various standard configurations, as shown in Figure 3.1.

A solar photovoltaic (PV) water pumping system consists of the following main components: the PV array, with support structure, wiring and electrical controls; the electric motor; the pump; and the delivery system, including pipework and storage. These components have to be designed to operate together to maximize the overall efficiency of the system (or, rather, to optimize the cost-effectiveness of the system). An electrical controller is sometimes incorporated to improve the electrical performance of the system. Energy storage in the form of batteries is rarely used, as it is generally cheaper and simpler to store the water to cover periods of low solar input or high demand.

The advantages and disadvantages of the various pumping techniques are compared in Table 3.1. The main problem with PV pumps has been their high initial cost, but with cheaper PV modules coming onto the market and with improved system designs incorporating volume-produced pumpsets this does not constitute such a barrier.

Water supply

An example of a solar photovoltaic borehole pump used for village water supply is shown in Figure 3.2. Water supply requirements do not vary much month by month. It is important to provide sufficient storage to cover periods of cloudy weather, when the output from the PV pump will be low. A covered tank at or near ground level, connected to a number of automatic shut-off supply taps on a concrete or stone pad would be a typical arrangement.

Whenever possible, it is safer to take water intended for human consumption from closed boreholes or protected wells. If a surface water source, such as a lake or a stream, has to be used, it is usually possible to construct some form of filter

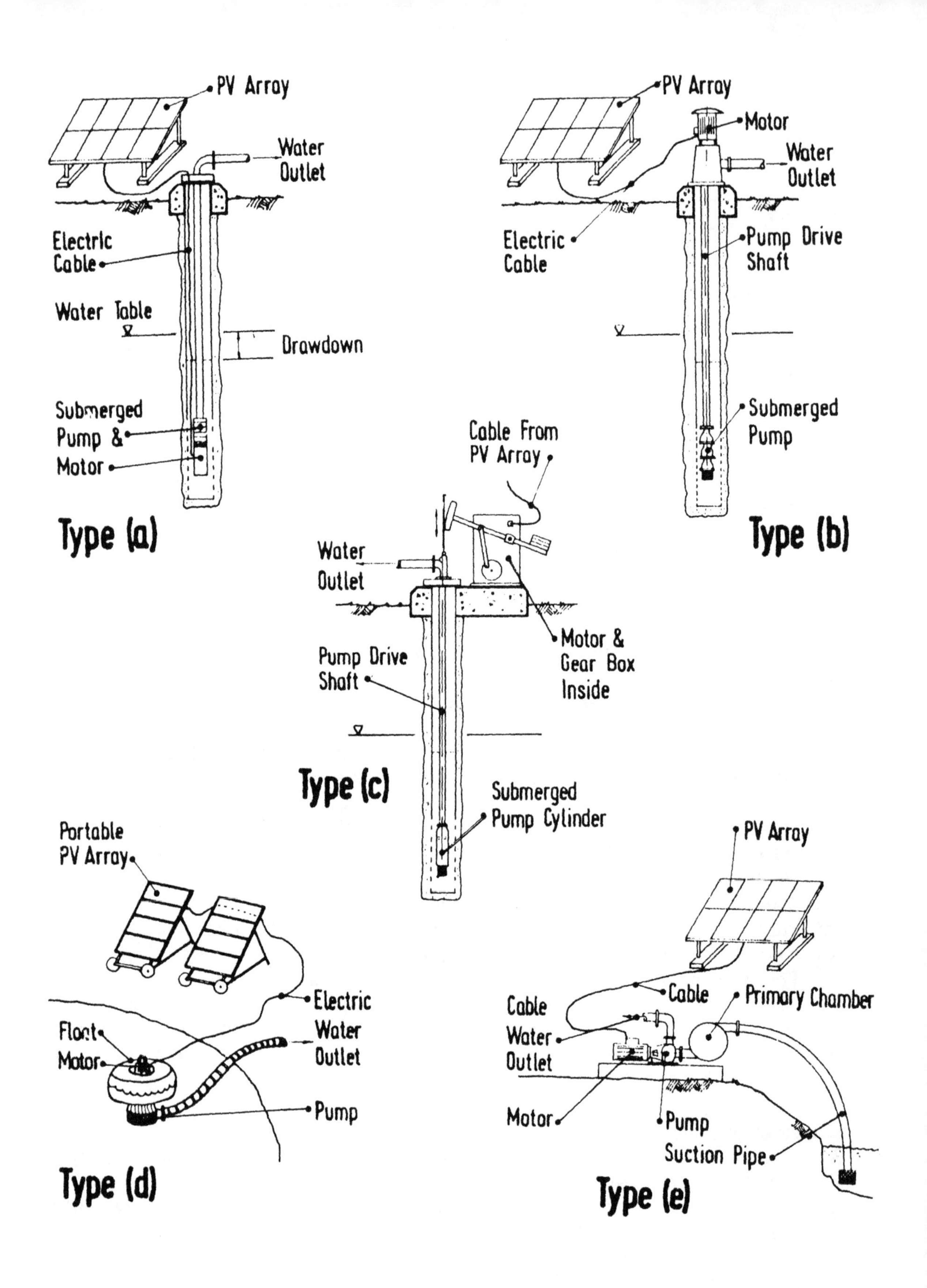

Figure 3.1 Standard Configurations of PV Pumps

Pumping technique	Main advantage	Main disadvantages
Hand pumps	Low cost Simple technology Easy maintenance	Low flow Absorbs time and energy that could be used more productively elsewhere. Often involves uneconomic use of expensive boreholes
Diesel and gasoline pumps	Low capital cost Can be portable Extensive experience Easy to install Easy to use	Maintenance often inadequate, reducing life. Fuel often expensive and supply unreliable. Noise, dirt and fume problems. Unreliable if not maintained
Wind pumps	Moderate capital cost Suitable for local manufacture Easy to maintain Non-polluting Need no fuel Long life Extensive experience	Very sensitive to wind speed, with periods of low output. Needs open terrain Not easy to install
Solar PV pumps	Low maintenance Non-polluting Need no fuel Reliable Long life System is modular	High initial cost Low output in cloudy weather

Table 3.1 Comparison of Pumping Techniques

when building the pump sump. Complete solar-powered water treatment plants are now available, as discussed in Chapter 7 of this survey.

Water intended for livestock is usually pumped from a borehole and stored in a raised tank so that the cattle drinking troughs may be gravity fed through ball valves. The PV array needs to be well-protected to prevent damage by livestock.

Irrigation

PV pumps are well suited to irrigation applications. They produce the most water when the solar radiation is greatest and hence when the crop water demand is highest. Because PV pumps deliver water over a period of about 10 hours each day, it is important to plan carefully the distribution of the water to avoid losses by evaporation and infiltration. The irrigation technique will need to be adapted to take best advantage of the available water. For example, instead of one large pump, it may be better to deploy several small pumps at different places in a large field, or in several separate fields. Alternatively, it may be feasible to store water for discharge at a higher flow rate over a shorter time.

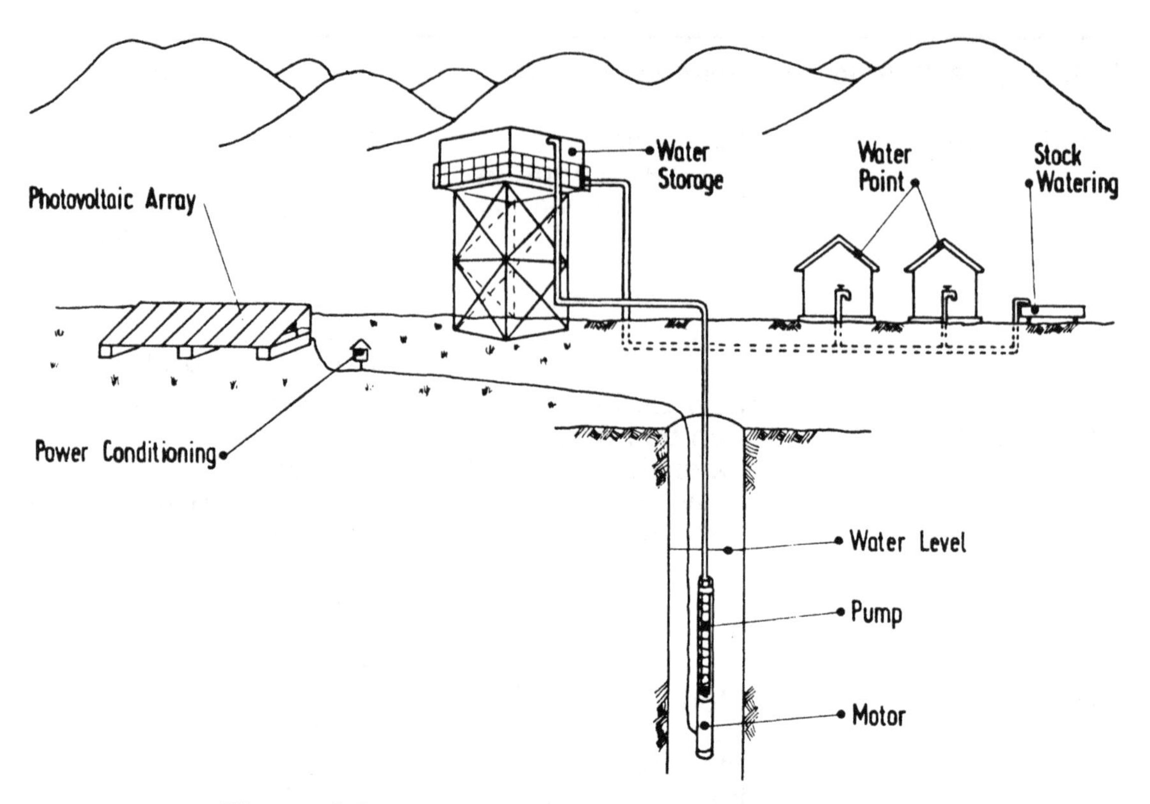

Figure 3.2 Typical PV Water Pumping System

The pumping of relatively large volumes of water for flood irrigation for rice is unlikely to be cost-effective, whereas a fruit farmer may well find that the relatively small volumes required for a trickle system could be supplied very economically. The underlying principle is that the cost of water used must be less than the value of the extra crop gained through the irrigation.

A typical PV irrigation pumping system is shown in Figure 3.3.

Field experience

A substantial volume of field experience is now available relating to solar pumps. Approximately 2000 units made up of the different configurations shown in Figure 3.1 have been supplied worldwide. A comprehensive study of solar pumps, involving field and laboratory testing of component and complete systems, was completed in 1983 by consultants for the UNDP and World Bank (Ref. 3.1). A survey of solar pumping field performance was recently been carried out by consultants for the World Bank (Ref. 3.2). There is also nearly ten years of experience available with Mali Aqua Viva, who have sponsored some 50 PV pumps, mainly for village water supply (Ref. 3.3).

3.2 Principal case study — Mali

Over the past ten years, more than 80 PV pumps have been installed in Mali, mainly under the auspices of the charitable organization Mali Aqua Viva (MAV) with financial support from various aid agencies. Other systems have been installed by various organizations, in particular four systems for which good data exist installed by LESO (Laboratoire de l'Energie Solaire), Bamako.

Information on the PV pumping experience in Mali has been obtained from personal knowledge supplemented by published data (References 3.3, 3.4, 3.5).

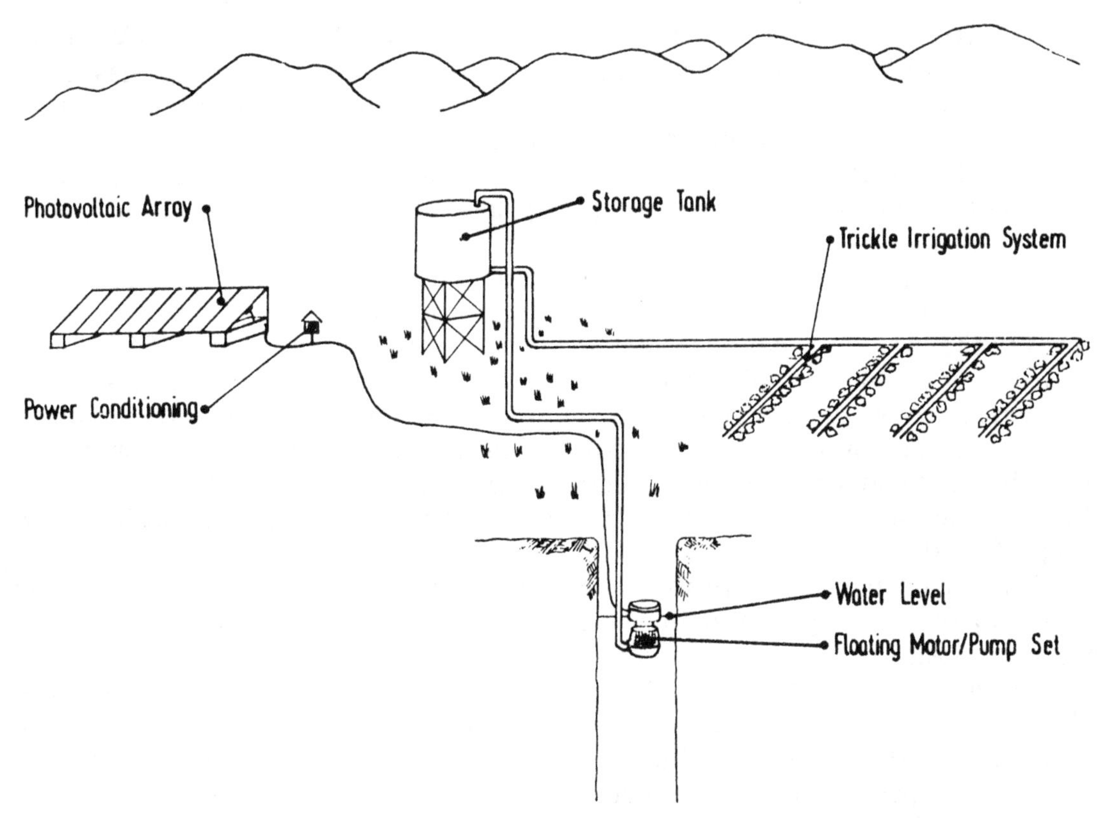

Figure 3.3 Typical PV Water Pumping System for Irrigation

Most of the systems installed have been for village water supplies and are well-appreciated by the users. Arrangements for technical support have been established, so that on-going advice can be given and faults corrected. The MAV approach to all their water supply improvement projects is to involve the local people from the beginning, to ensure their full understanding and commitment. In the case of solar pumps, the villagers are expected to build as much of the local infrastructure as possible (eg. storage tanks, access, foundations) and this means that a significant proportion of the total capital cost (up to 25 per cent) is met from local sources. The motivation generated by this initial involvement has proved to be a key factor in the successful implementation of most of the MAV projects.

Technical aspects

There are many practical difficulties in measuring accurately the performance of solar pumps in the field. The first attempt to do this in a systematic way was part of the programme of work included in the UNDP/World Bank solar pumping project, carried out between 1979 and 1983. Four PV pumps were installed and tested in Mali as part of this project and the results are presented in Reference 3.6.

More recently , LESO have been systematically monitoring the performance of a number of PV pumps and keeping detailed records of faults and other incidents that affect the system reliability. Staff from LESO have also recently visited over 30 PV pump sites in Mali, making notes on the technical condition of the system, the financial arrangements, operation and maintenance, and the social and institutional aspects (Ref. 3.4).

The experience in Mali is mainly with low and medium head systems incorporating various types of centrifugal pump. There is at least one surface-mounted positive displacement pump, but no submerged reciprocating pump (jack pumps, Type C in Figure 3.1). There have been few problems with the PV arrays, although discolouration of cells and corrosion of module frames has been observed. Most technical problems have been associated with the pumps, motors and control systems. In some cases, the well itself was not vertical or not deep enough, which lead to major pump problems. In other cases the yield of the well was insufficient to match the pump capacity, leading to the pump running dry with consequent failure.

Originally, the majority of the PV pumping systems installed in Mali had submerged pumps and surface mounted DC motors (Type B in Figure 3.1). More recently the systems have been of the AC submersed pumpset type (Type A in Figure 3.1), a change which has increased reliability and eliminated a major maintenance cost, namely shaft repair. In the multi-stage vertical turbine pumps, vibration in the connecting drive shaft has caused at least one broken shaft and many bearing and seal failures. Directly-coupled motor/pump units, either down the borehole for medium and high head applications or floating units for low head are preferred.

There have been a few problems with motors, principally due to overheating caused by overloading. The main electrical failures have been with the electronic control systems associated with some systems.

The availability (proportion of time operation) of the systems installed prior to 1984 has been generally found to be between 70 to 85 per cent, due largely to the delays involved in getting faults repaired. The availability of the more recent systems is expected to remain much higher, at around 95 per cent.

Where it has been possible to measure the actual output of water in relation to solar energy input, it has often been found that manufacturers' claims have not been met. However, this result is not uncommon for many commercial systems, including conventional diesel-powered pumps. It underlines the need for clear specifications at the tender stage and for independently-witnessed performance tests before shipment to the site.

The independently measured performance of four solar pumps in Mali is shown in Figure 3.4 (from Reference 3.7).

Economic aspects

The organizations installing PV pumps in Mali have found that capital costs have been steadily falling over recent years. A 1300 Wp system purchased in 1979 cost \$35/Wp, but now a similar system would cost about \$11/Wp. To this must be added a further \$6-\$10/Wp for shipping and installation.

Most of the PV pumps in Mali are administered by a village co-operative, which in addition to organizing the local support for the initial installation, also arranges for water charges to be levied on users. Some well-organized co-operatives are able to raise enough money in this way to pay for on-going costs and for further development, such as a second pump. In other places, the organization is not so efficient or the circumstances may not be favourable, leading to insufficient local funds to pay for maintenance and any further development.

A recent study of solar system economics in Mali found that PV pumps could be competitive with diesel pumps especially for low lift applications (Ref. 3.5). An example of a typical economic analysis for a 1400 Wp low lift system is shown in Table 3.2. Based on a 15 year period and 10 per cent discount rate, the annual levelized cost comes to \$3905. If 100 per cent of the water output is useful, the average unit cost is \$0.07/m^3. Even if it were only practicable to charge for 25 per cent of the water produced by the system, at say the equivalent in local currency of \$0.25/m^3, the system would still be able to compete with a diesel pump.

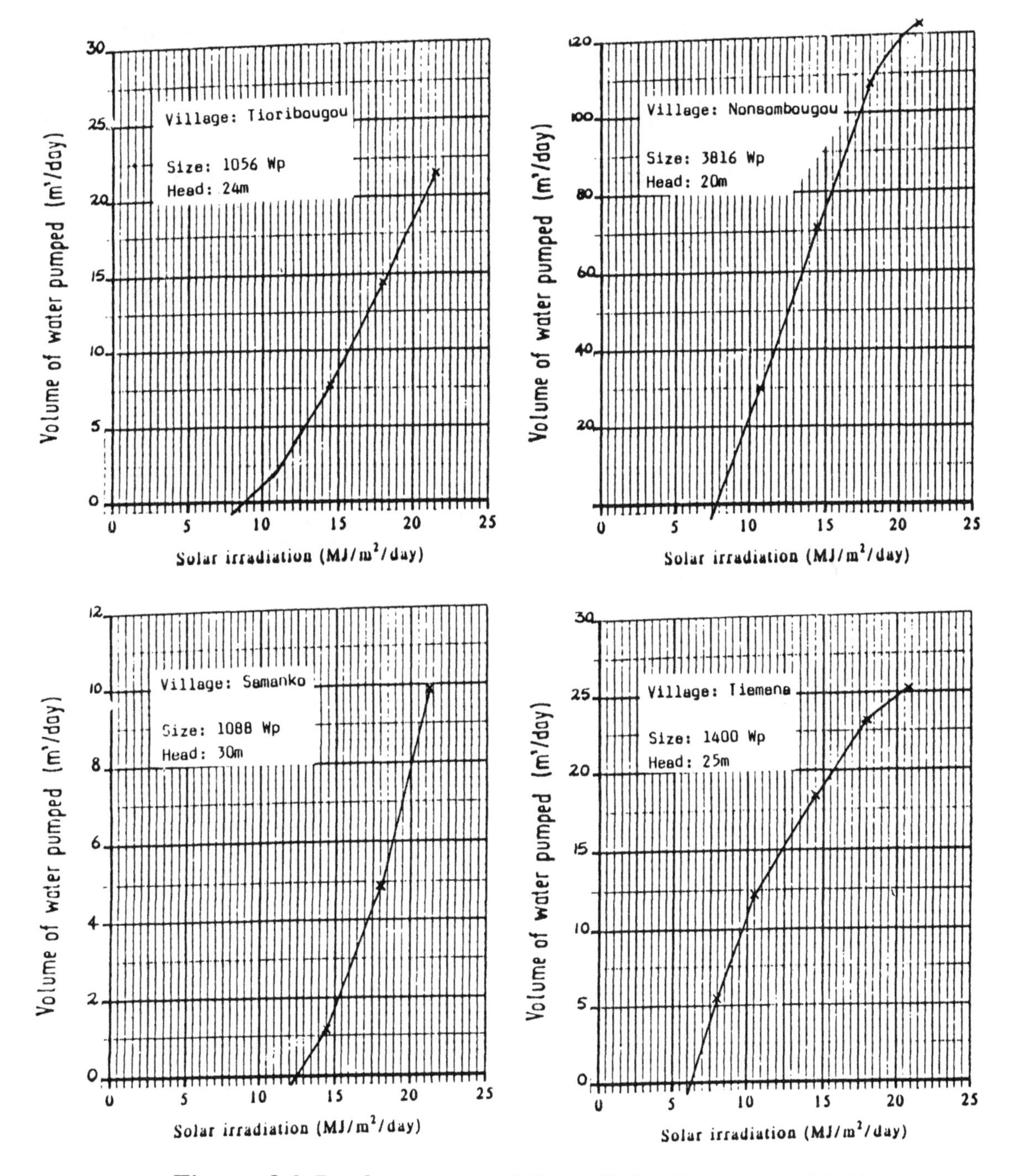

Figure 3.4 Performance of Four Solar Pumps in Mali

Social and institutional aspects

The PV pumping systems have generally been well-received in Mali, particularly at the sites where there has been a good history of local involvement with the installation from the planning stage. Demand is usually greater than supply and this is the main reason for user dissatisfaction. There have also been problems with pipework, storage tanks and water distribution arrangements, which some users have perceived as being failures of the PV system.

Because of organizations such as MAV and LESO, maintenance and technical support for PV pumps is probably better in Mali than anywhere else in the developing world. Nevertheless, there have been problems in repairing faults, due to poor communications. In many cases, the overseas supplier of a system has not responded promptly to request to repair faults or to supply spare parts, which is disappointing. Efforts are now in hand at LESO to train engineers and technicians in PV system technology and to carry out trouble-shooting, repairs and routine maintenance.

The level of technical support required may be judged from the cost of maintenance. According to figures given by MAV, it has cost on average about $540 per pump per year to maintain some 30 PV pumping systems. With the experience that has been gained and the introduction of improved system configurations, the average cost of maintenance is expected in future to come down to around $330 per pump per year. LESO has estimated the cost of maintaining the submerged AC motor/pump systems (Type A in Figure 3.1) to be $250 per pump per year.

BASIC DATA	
System size	1400 Wp
Capacity	160 m³/day at 5m head
Hydraulic power	361 W
Use (assumed availability)	340 days/year
Annual output	54 400 m³
COST ANALYSIS	
Period of analysis	15 years
Discount rate	10%
Capital costs:	
— equipment CIF	$16 052
— installation	$ 8 250
— total	$24 302
Replacement costs, 2 pumps	£ 3 500
Total Present Worth of system	$27 802
Recurrent costs:	
— maintenance	$ 250 per year
— Present Worth	$ 1 902
Total Present Worth of Life Cycle Costs	$29 704
Annual Levelized Cost	$ 3 905
Units Water Cost:	
— 100% use	$0.07 per m³
— 50% use	$0.12 per m³
— 25% use	$0.29 per m³

Table 3.2 Typical Cost Analysis of PV Pumping System in Mali

3.3 Other water pumping projects

UNDP/World Bank project

Reference has already been made to the comprehensive evaluation of solar pumping technology and economics carried out by consultants for the UNDP/World Bank over the period 1979-83. The project also resulted in the production of a solar pumping handbook covering the technology and economics and giving recommendations on procurement of appropriate equipment (Ref. 3.8).

In addition to providing much practical advice on all technical aspects of PV pumping systems, the study also presented general guidance on the economic prospects. Based on current costs for installed systems in the range $15-23/Wp, it was found that for irrigation applications, solar pumps could be cost competitive with diesel pumps for small-scale, low-lift applications (eg. 30m^3/day through 5m head) in places where the average solar input was at least 15 MJ/m^2 per day (4.2 kWh/m^2 per day). For water supply applications, solar pumps could be competitive with diesel pumps for relatively low-flow, medium head applications (eg. 25m^3/day through 10m head).

India

India has been manufacturing PV cells and systems using practically 100 per cent local components since 1979. A number of solar pumps have been installed at various villages in India for water supply and irrigation applications. In many cases, the technical aspects have been completely overshadowed by the social and institutional problems encountered.

In one case, a PV pumping system was chosen because of a constant history of breakdowns and fuel supply problems for an existing diesel pump (Ref. 3.9). The water depth was 15m, too deep for animal-powered devices. Construction of the well and installation of the pump was subjected to many delays and a solar management committee came into being as the most acceptable means of ensuring the equitable distribution of the water produced by the pump.

Botswana

The results of a study are reported in References 3.10 and 3.11 on the economics of PV pumps in comparison with diesel pumps for rural water supplies in Botswana. The studies compared a 6 kW diesel engine powering a progressing cavity pump with a PV-powered electric motor driving the same pump. The study found that a significant technical consideration was the borehole yield characteristics. In a typical application, a maximum flow of 2 to 6 m^3/hour will occur with the PV system around noon on clear days. Yield tests indicated that this would exceed the rate of water inflow into the well, resulting in the well level falling too low.

Another relevant issue concerns the use of existing pumps and wells to minimize the cost of establishing a PV system. This requires good communication between the field personnel and the equipment suppliers to ensure a properly designed system. Economic analysis showed that the PV option would be cost-competitive with the diesel option at current prices.

Egypt

Two PV pumps have been operating since 1981 as part of the Desert Development Demonstration and Training Project at Sadat City (Ref. 3.12). A 10 kWp array coupled to an inverter supplies AC power at 220 V, 50 Hz to the headquarters buildings and to an AC submersible pump set in a deep tubewell (43m head). A separate 3 kW system provides DC power to a surface-mounted motor driving a submerged progressing cavity pump. These is also a surface-mounted booster pump for irrigation.

The PV arrays have operated reliably with average daily conversion efficiency reported as 7.22 per cent. The AC submersible pump has also performed well. The progressing cavity pump has run reliably since mid-1984. Prior to that, excessive vibration of the drive shaft reduced the performance of the pump and resulted in a number of failures. The problem was solved by installing additional guide bearings to the shaft, since when the pump has operated reliably. The AC system includes batteries which need regular maintenance.

3.4 Conclusions

Technical aspects

PV pumping technology has improved significantly over recent years, with the emphasis on better matching of system components, increased reliability and reduced maintenance requirements. The type and size of system needs to be chosen carefully on the basis of a systems approach to the problem, taking into account all relevant factors, including operation and maintenance implications. (See Reference 3.8 for design and procurement recommendations).

PV arrays based on the well-proven crystalline silicon cell technology have generally proved to be the most reliable component of a pumping system. With the advent of lower-cost thin film cell technologies, it will be necessary to monitor array performance carefully to ensure adequate provision is made for long-term degradation.

The introduction of brushless DC motors by some manufacturers for surface mounted or floating pumps has eliminated the need for brush replacement. For submerged motors, water-filled AC induction motors are proving to be much more reliable than sealed DC motors. This type of borehole system is now preferred to turbine pumps with surface motors and long vertical drive shafts. The variable frequency DC-to-AC inverters required for AC systems provide a low-cost means of matching the PV array output to the motor load and they appear to perform reliabily in the field.

Centrifugal pumps can be well-matched to PV arrays. Problems have been reported with surface-mounted centrifugal pumps, due to the need to maintain prime. A self-priming tank on the suction side has proved to be more reliable than a foot valve. Centrifugal pumps should not be used for suction lifts more than 5 to 6 m and wherever possible a floating or fully submersed unit is to be preferred.

Positive displacement pumps have a water output that is practically independent of head and directly proportional to speed. For PV-powered systems, problems have been experienced due to the cyclical nature of the load on the motor and the high frictional forces, particularly at start up. At high heads this type of pump can be more efficient than a centrifugal pump, since the frictional forces are relatively small compared with the hydrostatic forces. Positive displacement pumps are however usually very rugged and reliable, provided the overall system has been well-designed in the first place to suit the conditions obtaining at the site.

A common cause for pump and/or motor failure has been overloading due to sediments in the water or tight shaft bearings. Dry running due to loss of prime (surface pumps) or falling water level in the well is another common cause of failure. Increasingly manufacturers are providing low water level and/or high temperature protection for the motor.

The use of tracking PV arrays, maximum power point trackers and batteries may offer advantages in theory, but experience has shown that, at remote sites where maintenance is difficult to provide, the extra complexity introduced is counter-productive.

Pump performance is heavily dependent on the assumptions made at the design stage regarding solar and water resource characteristics. Careful account has to be taken of the variations in solar input to the array, the static water level in the well and the water demand. Failure to do this has resulted in many systems being undersized so that they fail to meet the demand, or excessively oversized, with associated additional capital cost. However, these problems should diminish as experience builds up and a larger data base applicable to each country becomes available to system designers and suppliers.

Economic aspects

The F.O.B. prices of PV pumping systems have been steadily falling from about $30/Wp in 1978 to as low as $10/Wp in 1986. To this has to be added the cost of shipping and installation. The unit water costs expressed per volume-head products ($/m^4) may be calculated on a life cycle cost analysis for different assumptions regarding demand, solar insolation and fuel costs. This has been done in Reference 3.13 the results of which we presented in Figure 3.5. The data assumed in the analysis is given in Table 3.3. It can be seen that solar pumps are typically competitive up to 1000 m^4/day demand (eg. 40 m^3/day pumped 25m. This approximates to 1400 Wp array power. Reference 3.2 also concluded that PV pumps are competitive up to approximately 1000 Wp (and for regions of very high diesel operating cost up to 2.5 kWp).

Water for irrigation is characterized by a large variation in demand from month to month. Hence a solar pump sized to meet the peak demand is under-utilized in other months. This adversely affects the economics/unit water costs and it is for this reason that solar pumps are more competitive for rural water supply.

Social and institutional aspects

PV pumping systems, being a new technology, need continuing institutional support to enable them to be successfully integrated into the rural communities that stand to benefit.

There are three main areas where institutional support is particularly needed:

— at the planning and procurement stage
— for administering the operation
— maintenance and spare parts.

At the planning stage, it is important to involve the local community from the outset and encourage them to organize a management committee. The local costs should be raised locally, either in cash or in direct labour. Clearly experienced technical advice will be needed for the design and procurement of suitable equipment.

The local organization must then be assisted to organize appropriate arrangements for distributing the water and levying charges. It is helpful to appoint a keeper or operator to watch over the system and he will need to be given basic training in routine maintenance and simple trouble-shooting.

With good design and the installation of the latest types of system, system reliability should be good. However, there will inevitably be faults arising from time to time which cannot be fixed by the users. Established arrangements need to exist for calling in technical support and for the procurement of any necessary spare parts. Good communications between the site and the source of support are of course very desirable, but it has to be recognized that in many rural areas the necessary infrastructure simply is not available.

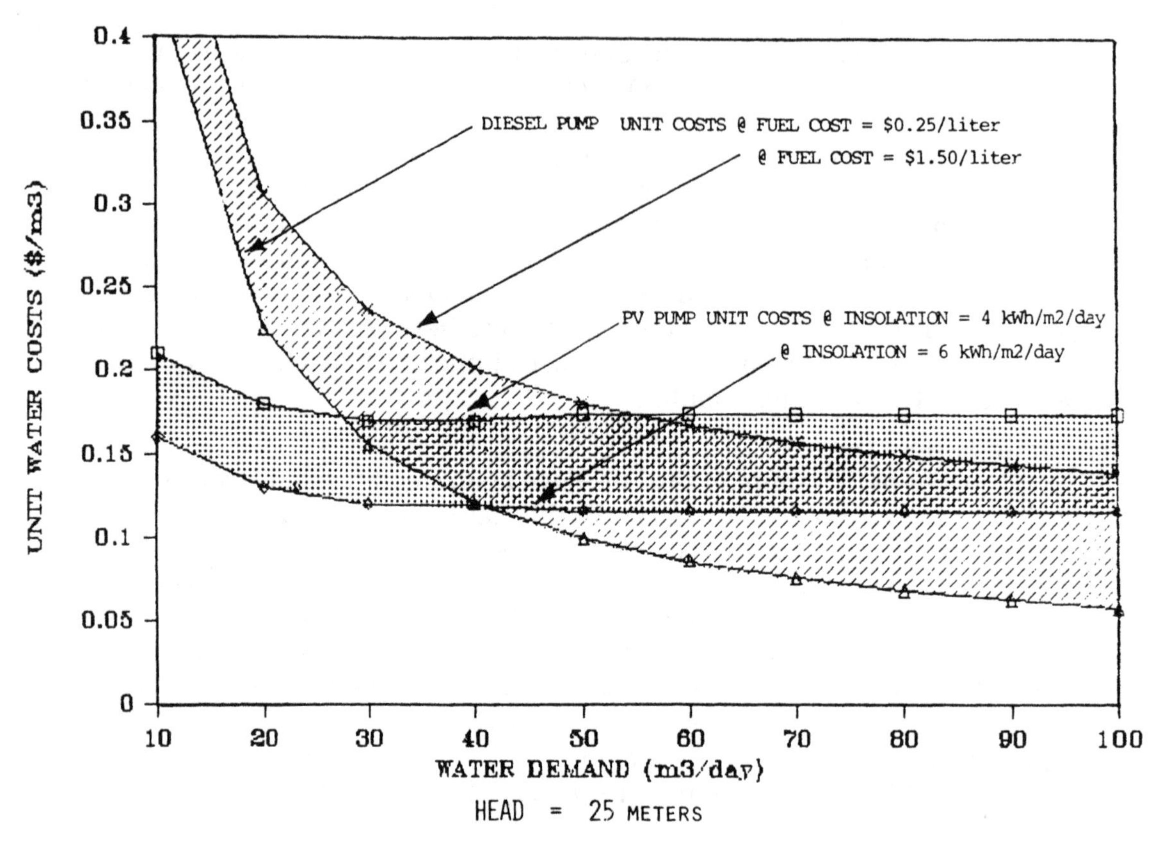

Figure 3.5 Solar Pump and Diesel Pump Cost Comparison

PV PUMP SYSTEM CHARACTERISTICS	
Depth of Water Supply, Head	25 m
Annual Average Daily Water Demand	20 m³/day
Annual Max Daily Water Demand	30 m³/day
Insolation	5.0 kWh/m²/day
PV Array Peak Power	1.11 kWp
PV Pumping System Capital Cost	$12.00 $/Wp
PV Pumping System Availability	95.0 %
PV Array Life	20.0 yrs
Pump Life	5.0 yrs
Normal Discount Rate	10.0 %
Inflation Rate	5.0 %
PV NPV Unit Costs	$0.14 /m³
DIESEL PUMP CHARACTERISTICS	
Depth of Water Supply, Head	25 m
Annual Average Daily Water Demand	20 m³/day
Annual Max Daily Water Demand	30 m³/day
Diesel Generator Power Rating	3.0 kW
Average Load Factor	45.4 %
Diesel Fuel Cost	$0.75 /litre
Diesel Gen-Set Pump Capital Cost	$1.59 /W
Diesel Pump Availability	90.0 %
Diesel Gen-Set Life	6.0 yrs
Pump Life	5.0 yrs
Nominal Discount Rate	10.0 %
Inflation Rate	5.0 %
Diesel NPV Unit Costs	$0.26 /m³

Table 3.3 Data Used for Solar Pumping Cost Comparison

Chapter 3 — References

3.1 'Small-scale solar-powered pumping systems: the technology, its economics and advancement'. Main report prepared by Sir William Halcrow and Partners in association with I.T. Power Ltd, for UNDP Project GLO/80/003 executed by the World Bank, June 1983.
3.2 IT Power, 'Solar-Powered Pumping Systems: Their Performance, and Economics'. Report to World Bank (1986).
3.3 Mali Aqua Viva Report No. 7, Activities from October 1979 to October 1981.
3.4 Private communication from Laboratoire de Energie Solaire, Bamako, Mali, Sept 1985.
3.5 T.R. Miles Jr, 'Economic Evaluation of Renewable Energy Techologies at LESO' for US-AID and Energy/Development International, Dec 1985.
3.6 'Small-scale solar-powered irrigation pumping systems — Phase 1 Project Report' by Sir William Halcrow and Partners in association with Intermediate Technology Development Group Ltd, for UNDP/World Bank, July 1981.
3.7 'Small-scale solar-powered irrigation pumping systems — Technical and Economic Review' by Sir William Halcrow and Partners in association with Intermediate Technology Development Group Ltd, for UNDP/World Bank, Sept 1981.
3.8 J.P. Kenna and W.B. Gillet, *Solar Water Pumping — A Handbook* IT Publications, London, 1985. 1985.
3.9 P. Amado and D. Blamont, 'Implementation of a solar pump in a remote village in India; economical and socio-economic consequences — three years of working experience', *Proc. of Third International Conference on Energy for Rural and Island Communities*, Inverness, UK, Sept 1983.
3.10 R. McGown and A. Burrill, 'Current Developments in Photovoltaic Irrigation in the Developing World', A.R.D. Inc., 1985.
3.11 D. R. Darley, 'PV vs Diesel: A Grounded Economic Study of Water Pumping Options from Botswana', Massachussetts Institute of Technology, USA, 1984.
3.12 IT Power Inc, Proceedings of the Photovoltaics Information, Symposium and Workshops, Nairobi and Chiang Mai, April 1986. (Sponsored by the UNDP Energy Office).

4. PHOTOVOLTAIC REFRIGERATORS FOR RURAL HEALTH CARE

4.1 Introduction

The need for PV refrigerators

In many developing countries, living conditions for the majority of the rural population are poor and there is widespread malnutrition combined with a high incidence of disease. Infant mortality is particularly high in the rural areas, where in some countries, as many as one third of the children die before the age of two. Much of the disease could be eliminated or controlled through mass immunization the practical problems involved are formidable. Most countries are however making large efforts to improve the quality of rural health care, including expansion of their immunization programmes.

Vaccines require refrigeration during transportation and storage to remain effective. It is important to maintain the vaccine 'cold chain' from the place of manufacture right through to the point of use. This imposes a major logistical problem because generally there are no reliable electricity supplies to operate conventional electric refrigerators in the rural areas where the clinics and health centres are located. Kerosene and bottled gas (LPG) powered refrigerators are available but their performance in many cases is not adequate and there are often problems in ensuring regular fuel supplies.

Solar photovoltaic refrigerators have the potential for better performance, lower running costs, greater reliability and longer working life than kerosene or LPG refrigerators, or diesel generators powering electric refrigerators. Recognizing this potential, the World Health Organisation (WHO), the Centre for Disease Control (CDC), the US Agency for International Development (US-AID), the European Community (EC) and other agencies have installed and evaluated many PV refrigerators throughout the developing world. At least 800 PV medical refrigerators have been installed over the last five years, mainly for testing and/or demonstration purposes. The stage has now been reached where a number of designs have been approved by the WHO, opening the way to wider implementation of this technology.

The use of PV refrigerators instead of kerosene or LPG units offers the following benefits;

i) Elimination of fuel supply costs and delivery problems
ii) Reduced vaccine losses through improved refrigerator reliability, with associated reduced anxiety among medical personnel
iii) Reduced maintenance workload for technicians and medical personnel, with associated cost and time savings
iv) Overall cost savings for the vaccine cold chain equipment.
v) A more effective and sustainable immunization programme, leading to reduced incidence of disease.

The technology

Five alternative methods of solar powered refrigeration were surveyed by the WHO during 1980 (Ref. 4.1). These were: photovoltaic/vapour compression, photovoltaic/Peltier effect, solid absorption/zeolite, solid absorption/calcium chloride, and liquid absorption/ammonia. Photovoltaic systems are the only type commercially available for vaccine storage (with the exception of one solid absorption ice-making plant manufactured in Denmark). Of the photovoltaic systems, several manufacturers offer vapour compression systems in suitable forms for use in the vaccine cold chains.

A schematic diagram of a photovoltaic/vapour compression refrigerator is given in Figure 4.1. A PV array charges a battery via a charge regulator, to ensure that the battery is not overcharged. The battery powers a DC motor which is coupled directly to the compressor. The motor/compressor is usually manufactured as a hermetically sealed unit. The motor is of the electronically commutated brushless type. A second regulator is employed to ensure that the motor/compressor operates only within its rated power range and to prevent over-discharge of the battery. Freon refrigerant is used in the cooling cycle in the normal way, ie. the cooling effect is achieved by the heat absorbed by the refrigerant as it evaporates in the evaporator. A thermostat switches the motor/compressor unit on and off as required. Some models have two compressors and thermostats, one each for the refrigerator compartment and the freezer compartment.

The insulation is normally of the expanded polyurethane type and double the usual thickness to reduce heat gain and thereby reduce the energy consumption and increase the time the refrigerators can maintain safe temperatures with no power. Most units are top opening, to reduce loss of cold air and often have a secondary hinged or removable cover under the main door. Figure 4.2 shows a typical example of a PV vaccine refrigerator.

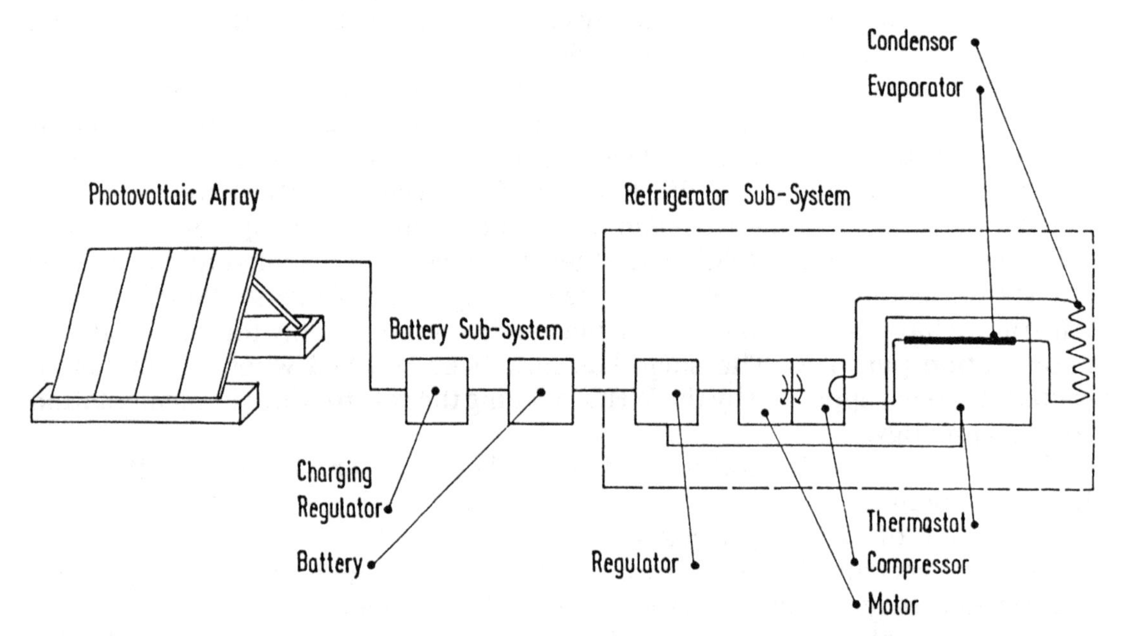

Figure 4.1 Schematic of Photovoltaic Refrigerator System

WHO specification

The vaccine capacities of solar refrigerators available or being developed vary widely, from 3.6 to 200 litres. The need for solar refrigerators is greatest at the peripheral health centres serving populations of 20,000 to 100,000. The quantity of packed vaccine needed to immunize fully 150 infants and their mothers is approximately 4 litres. There is however no general agreement yet on the best size

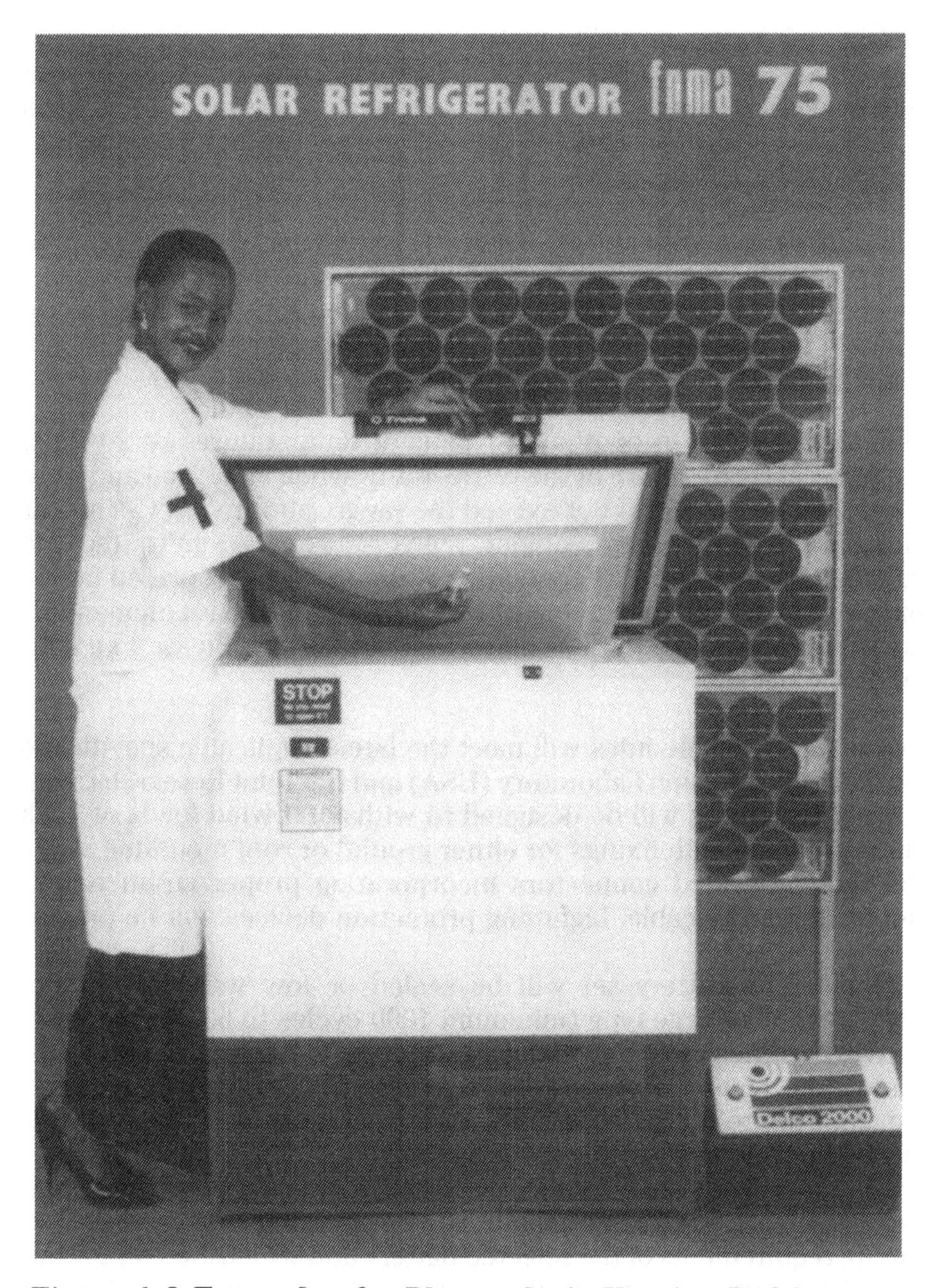

Figure 4.2 Example of a Photovoltaic Vaccine Refrigerator

for a PV vaccine refrigerator. Opinions differ on the quantity and volume of other biological products (eg. blood) which might be stored in the health centre refrigerator and many people believe that a larger cabinet will have a wider market. It is also important for the system to have the capacity to freeze ice-packs which are used when transporting vaccine from the health centre for immunization in the field. The ice production capacity is a significant load on the system and has a major influence on the PV array size and hence system cost.

In 1981, the WHO issued an outline specification for PV refrigerators which laid down minimum requirements covering vaccine capacity, ice-making performance, refrigerator performance, hold-over time, battery maintenance interval, etc. The basic requirements have been modified in the light of field experience and the current WHO specification (Ref. 4.2) provides as follows:

1. The design of the system will be sized to enable continuous operation of the refrigerator and freezer (loaded and including icepack freezing) during the lowest periods of insolation in the year. If other loads, such as lighting, are included in the system, they should operate from a separate battery set, not from the battery set which supplies the refrigerator.

2. The design of the system will permit a minimum of five days continuous operation when the battery set is fully charged and the photovolatic array is disconnected. During this time the internal temperature of the refrigerator will remain within the range of +0 to +8°C when the constant external temperature is a minimum of +32°C.

3. Refrigerator/Freezer: In continuous ambient temperatures of 20°C, 32°C and 43°C, the internal temperature of the refrigerator, when stabilized and fully loaded with empty vaccine vials, will not exceed the range +0°C to +8°C. This range will be maintained when, in an ambient temperature of +22°C, the maximum recommended load of icepacks containing water at +22°C is placed in the freezer and frozen solid without adjustment of the thermostat. The recommended load of icepacks will freeze in less than 24 hours and will weigh at least 2 kg without the material of the pack.

4. Photovoltaic Array: Modules will meet the latest applicable specifications laid down by the Jet Propulsion Laboratory (USA) and the Joint Research Centre, Ispra (Italy). Array structures will be designed to withstand wind loads of +200 kg/m^2 and will be provided with fixings for either ground or roof mounting. Appropriate photovoltaic-type sealed connectors incorporating proper strain relief will be provided for the array cable. Lightning protection devices will be provided.

5. Battery Set: The battery set will be sealed or low waterloss or non-liquid electrolyte deep discharge type (minimum 1000 cycles to 50 per cent discharge). Automotive batteries are specifically unacceptable for this application. The batteries will be housed within the refrigerator freezer cabinet, or in a cabinet separate from the refrigerator, but in either case lockable. No dry cell batteries shall be used to power instruments and controls.

6. Voltage Regulator: A voltage regulator will be provided, which meets the charge/temperature requirements of the selected battery and which cuts off the load when the battery has reached a state of charge which can be repeated to a minimum of 1000 cycles. Lightning protection will be provided. The load should be automatically reconnected when the system voltage recovers.

7. Instrumentation: A LED alarm will be installed to warn that power to the compressor has been cut by the regulator. An expanded scale voltmeter or a LED alarm will be installed to warn the user when the battery charge is in an unusually low state of charge to give adequate advance warning. The warning light of the minimum voltage limit should be clearly labelled 'DO NOT FREEZE ICEPACKS' in the appropriate local language. If an external reading thermometer is provided for the refrigerator, it should be marked clearly in green between 0°C and +8°C.

A thermostat or a defrost switch should be provided but no other power switches should be installed. Circuit breakers of cartridge fuse holders will be fitted with a polythene bag holding 10 spare fuses and special attention will be given to corrosion of fuse mountings.

8. Individual sea-crating of the components of each system should be provided whether or not containers are used to transport the systems. No package should be heavier than can be handled by hand in the country. Labels bearing handling instructions should be printed also in the appropriate local language.

9. Essential spare parts which may be needed during the first three years' operation should be assembled as a kit in appropriate quantities for central and regional storage in the country. A minimum list is as follows:

Item	*Quantity per 10 systems*
1. Photovoltaic modules	2
2. Regulator components (sets)	2
3. Battery sets	2
4. Array cables	1
5. Compressor, complete	1
6. Spare compressor regulator cards	3
7. Thermostat	3

10. Manuals will be provided for the installation and use of each system.

The WHO guidelines also require the supplier to provide a warranty for the replacement of any component which fails due to defective design, materials or workmanship. The minimum period of the warranty is required to be 10 years for the PV array, five years for the batteries and two years for the remaining components. The system supplier is also required to provide technical support for maintenance and repair operations in the country concerned for a period of at least two years. The supplier has to train an engineer to assist with installation of each system and also train users and repair technicians in each area.

Commercially available equipment

At least 20 companies now supply PV refrigerators for vaccine storage. To assist health authorities choose appropriate equipment, the WHO Expanded Programme on Immunization (EPI) publishes Product Information Sheets (Ref. 4.3). Inclusion of a product on these sheets in effect means that based on the information and test results available with WHO-EPI, the product is approved for use in the vaccine 'cold chain'.

The PV refrigerators currently approved by the WHO-EPI for use in the vaccine 'cold chain' are listed in Table 4.1. It should be noted that there are many 12 V DC-powered refrigerators available, intended primarily for leisure applications in boats and caravans. These have been designed for low capital cost, without consideration of energy consumption or internal temperature variation. Although such systems can readily be adapted for PV powering, they are not suitable for vaccine storage.

Field experience

The most significant work on testing and evaluating PV refrigerators in the field has been carried out under the WHO-EPI programme, involving field testing of over 50 systems for 12 suppliers in some 30 countries. Other work has been carried out by UNDP, UNICEF, European Development Fund, AFME (France), GTZ (W.Germany), ODA (UK), Oxfam and other aid agencies.

System Supplier	Refrigerator/Freezer Unit	Net Vaccine Capacity (litres)	
		Refrigerator	Freezer
AEG-Telefunken (W Germany)	Polar Products RR2	90	33
BP Solar (UK)	LEC EV5750	24	6
FNMA (Zaire)	FNMA 75	27	10
Leroy Somer (France)	Leroy Somer R50 + IF5	16	16
Polar Products (USA)	Polar Products RR2	90	33
Solarex (USA)	Polar Products RR2	90	33
Solarex (USA)	Marvel RTD4	80	10
Solavolt (USA)	Marvel RTD4	80	10
Solavolt (USA)	Polar Products RR2	90	33

Table 4.1 PV Refrigerators Approved by WHO-EPI for Vaccine Storage

Not all these projects have been consistently monitored, but two major projects are in progress which should provide substantial operating data in due course. Both these projects are funded by the European Development Fund, one involving the installation of 100 systems in Zaire and the other 20 systems in the South Pacific. A summary of experiences with photovoltaic refrigerators for medical use is given in Reference 4.4.

4.2 Principal case study — India

Background

Five PV refrigerators for vaccine storage and other medical uses in India have been selected for detailed monitoring, at the following locations;

- Dappar Subsidiary Health Centre (SHC), Patiala District, Punjab.
- Adalaj Public Health Centre (PHC), Gandhinagar District, Gujarat.
- Tirupparankundram PHC, Madurai District, Tamil Nadu.
- Solur PHC, Bangalore District, Karnataka.
- Balesar PHC, Jodhpur District, Rajasthan.

The PV refrigerators at these five sites were installed in August 1984. All five systems are similar and were supplied by AEG-Telefunken. They incorporate refrigerators/freezer units supplied by Electrolux. The PV array consists of 12 AEG-Telefunken type PQ 10/20/0 modules giving a total rated power of 230 Wp at 12 V. The regulator/control unit includes a shunt regulator to prevent overcharging and a load-disconnect switch to protect against over-discharging of the battery. The battery consists of two Varta 6 V units, with a total capacity of 150 Ah at 12 V. The refrigerator is an Electrolux type RCW 42 DC unit. The daily energy demand was estimated by the manufacturer to be between 670 and 860 Wh/day, depending on the site and climatic conditions.

Assessment of design

System performance assessments for each of the five systems in India were made for WHO. The assessments are based on the local climatic data (solar insolation and temperatures for each month) and calculated cooling load data. A 20°C rise above ambient was assumed for the normal operating cell temperature (NOCT) and an array power temperature coefficient was taken as 0.005. A day/night demand ratio of 0.67 and a battery turn-around (energy out to energy in) of 0.78 were assumed.

The analysis of the Tirupparankundram system indicated that there are two months (November and December) when the system Loss of Energy Probability (LOEP) or Loss of Load Probability (LOLP) is greater than 10 per cent. The system could be upgraded to give a LOEP or LOLPof 1 per cent or less by increasing the array size by 9 per cent (one module) and the battery size by a factor of 2.

The analysis for the Solar PHC installation indicates that there are four months (July, August, November and December) when the system of LOEP/LOLP is greater than 10 per cent. The Adalaj PHC site provides a severe test of the PV refrigerator system because of the higher ambient temperatures experience there, which results in lower array output. The system LOEP/LOLP is projected to be greater than 10 per cent in four months (July, August, December and January).

The Balesar PHC system is projected by the supplier to have a higher demand than the other sites. As a consequence, the LOEP/LOLP is projected to be in excess of 10 per cent in four months (August, November, December and January). The Dappar SHC installation appears to be the severest test of the system design because of the low isolation in the winter months and the projected high demand.

Field experience

It is too soon for the results of field monitoring of the five systems to be available. However, preliminary comments on the performance have been given to WHO. The Tirupparankundrum PHC system was reported to be functioning well several months after installation.

The main concern is with the batteries, which were considered to have too small a capacity. Also the batteries supplied are not 'low maintenance', requiring frequent topping-up by site personnel. Another concern was that the site personnel could relatively easily alter the temperature setting of the thermostat, which could result in freezing of the vaccines.

Bhoorbaral system

A PV refrigerator supplied by Solar Power Corporation (USA) was installed at the end of 1981 at Bhoorbaral near Delhi. The system has a 355 Wp array and a 630 Ah battery and incorporates an Adler-Barbour refrigerator. The system is reported to be performing well.

4.3 Other PV refrigeration projects

NASA Lewis Research Center

The NASA Lewis Research Center (LeRC), US-AID and the Center for Disease Control (CDC) signed an agreement to develop and test PV refrigerators for vaccine storage in 1979. Refrigerator systems were first laboratory tested under supervision by LeRC. Thirty-five systems are now being evaluated in 25 countries. Twenty-nine are stand-alone systems and 6 are refrigerator/freezer units installed in larger PV systems in US-AID development projects. A summary of the field trial programme is given in Table. 4.2.

From October 1981 to July 1984, the refrigerator/freezers in the NASA trials accumulated almost 500 system-months of operation. They are reported to have operated correctly, maintaining internal temperatures within the specified range for a total of 84 per cent of the time for the SPC system and 83 per cent for the SVI Polar Products and Marvel systems (Ref. 4.5). Although this is not an acceptable level of availability for vaccine storage, it is considerably better than achieved by many kerosene refrigerators.

A full record was kept of component failures during the trials. Most faulty equipment was discovered during commissioning tests. The most damaging of problems, and the most difficult to diagnose, concerned the power cable connector of the SPC system, which became electrically disconnected whilst appearing to be connected.

A number of systems experienced times when the internal temperature went outside specified limits either as a result of component failure or from improper use or incorrect system sizing. Specific reasons cited were:

— defective components (temperature controllers, thermostatically controlled air doors, voltage regulators)

— incorrect setting of thermostat

— excessive amounts of warm material being placed in the refrigerator/freezer

— shadowing of the array as a result of poor siting or not clearing trees.

Sufficient performance data for system optimization studies could not be collected because of the failure of many monitoring instruments. The two most important instruments, the pyranometers for measuring solar energy input and the ampere-hour meters for system energy consumption, were particularly failure prone. Enough data was collected to indicate that in many cases the percentage run time and hence energy consumption was higher than anticipated.

Some user reactions to the systems installed by NASA-LeRC have been very favourable. For example, the following comment is taken from the World Bank pre-investment study mission to The Gambia (Ref. 4.6):

The mission was told that the two solar refrigerators had performed quite satisfactorily since their installation, except for one occasion when the unit at Kaur had to be defrosted and disconnected for about three days. This was done on instructions from NASA-LeRC in order to fully recharge the batteries, after the data had indicated that the batteries were not staying at an adequate voltage. There are two possible reasons for inadequate voltage: first, there may have been only a minor fault either in the charging circuit or in one of the batteries, since all the primary components appeared to be functioning correctly; second, there may have been some abuse of the system, eg. through overloading the refrigerator by the operators with food and drinks in addition to the normal vaccines and ice-packs. The former fault could have been corrected easily if the attendant were properly trained, as is planned through the present project. The latter fault could have been prevented by equipping the refrigerators with a padlock, as indeed the kerosene refrigerators are currently equipped. The experience with solar refrigerators in The Gambia can therefore be considered successful.

In Ecuador, the users were so impressed by the 100 per cent reliability of their systems that the officials now plan to make the health post with the PV refrigerator a central storage facility for all vaccine in that area (Ref. 4.5).

World Health Organisation

To prepare for field trials, the WHO organized laboratory tests on 12 types of refrigerators suitable for PV powering at the Consumers Association in the UK during 1981-83. The objectives of the tests were:

(i) To determine independently the most suitable refrigerator models

(ii) To identify possible component or system failure modes prior to expending resources on implementing field trials.

Of the 12 models tested, 10 passed the laboratory testing, a number after being referred back to the manufacturer for modification.

COUNTRY	SYSTEM SUPPLIER (Refrigerator Model)				POWER SYSTEM		AGENCY	
	SOLAVOLT (Marvel)	SOLAVOLT (Polar Products)	S.P.C. (Adler Barbour)	SOLAREX (Adler Barbour)	ARRAY PEAK WATTS	BATTERY AMP-HOURS	DATE INSTALLED	RESPONSIBLE AGENCY
AFRICA								
Burkina Faso		1			200	315	2/84	AID
Gambia			2		297	630	1/83	CDC
Ivory Coast			2		330	630	2/83	CDC
		1			280	315	2/84	AID
Kenya				2	*	630	5/83	AID
Liberia			1		363	630	10/84	AID
Mali		1			200	315	2/84	AID
Zaire			1		330	525	2/83	AID
Zimbabwe			1		264	630	2/83	AID
				1	*	630	5/83	AID
AMERICAS					264	630	9/82	CDC
Colombia			1					
Dominican Republic			1		264	525	8/82	AID
Ecuador			1		264	630	9/82	AID
				1	*	630	3/83	AID
Guatemala			1		231	630	10/82	AID
Guyana			1		264	630	9/82	AID
				1	*	630	2/83	AID
Haiti			1		264	630	9/82	AID
Honduras	1				200	420	1/84	AID
Peru			1		231	525	10/82	CDC
St. Vincents/	1				200	420	1/84	AID
Grenadines		1			160	315	1/84	AID
NEAR EAST								
Jordan	1				160	420	6/84	AID
Morocco			1		330	525	10/83	AID
Tunisia		1			240	315	2/84	AID
	1				240	420	2/84	AID
			1		*	630	1/83	AID
ASIA								
India			1		330	630	10/81	CDC
Indonesia			2		297	630	4/82	AID
Maldives			1		264	630	5/82	CDC
Thailand	1				200	420	11/83	AID

*Part of larger photovoltaic installations.

Table 4.2 WHO NASA-LeRC — CDC — US-AID Field Trial Programme

Following the laboratory testing, the WHO has been co-ordinating the implementation of the field trials programme carried out directly and by a number of other organizations. The WHO-EPI office in Geneva, Switzerland, acts as a clearing house for enquiries and the collection of field trial data. It also arranges for the analysis of the data and publication of reports. Further laboratory testing of refrigerators is continuing with WHO sponsorship at testing facilities at Cali, Colombia.

The primary objective of the field trials is to evaluate the performance and reliability of PV refrigerators when used in the EPI under widely varying climatic conditions. The first installations in the WHO-EPI programme started in 1983 and are being conducted in co-operation with other agencies and the manufacturers. The field trials in progress are summarized in Table 4.3. The current status of the programme is presented in Reference 4.7.

The principal interest is in the recent installations of BP Solar/Lec systems undergoing trials in Kenya, Tanzania and Ghana, the AEG/Electrolux systems in India and the Solar Force/Leroy Somer system in the Yemen Republic.

The WHO-EPI field trials have not been underway as long as the NASA-LeRC trials and hence the data collected so far is very limited. Initial indications are that some problems have been experienced. For example, systems installed in the Philippines were found to have an undersized PV array. The BP Solar/Lec system was reported to be successfully meeting expectations during visits in July and November 1984 (Ref. 4.8). The BP Solar/Lec systems are instrumented with comprehensive data logging equipment.

Trust Territory of the Pacific Islands

In 1981 Motorola (USA) was awarded a contract to supply 24 PV refrigerators (and also lighting sytems) to dispensaries and schools in the Trust Territory of the Pacific Islands. Motorola assigned the contract to Solavolt. The dispensary systems consisted of a 200 Wp array, a battery and a WSR-12 refrigerator. The installations started in March 1982 when training was also given.

The systems were easy to install but during the initial training seminars, defects with the WSR refrigerators emerged. Two units both had defective thermostat cards and repeated problems with blown fuses were experienced. Ten of the 24 systems were found to be out of action one year later, of which eight were repairable.

The main problem with this project was the poor refrigerator reliability. The WSR unit is no longer commercially available.

Zaire

The world's largest programme to use PV refrigerators (100 systems) and also lighting systems (750 installations) is currently underway in Zaire (Ref. 4.9). Systems are being installed in clinics and dispensaries throughout the country. The programme is the responsibility of the Département de la Santé Publique and is being financed by the European Community (European Development Fund). The systems are supplied by Solarforce (France), working with a local company, FNMA, as sub-contractor.

The FNMA PV refrigerator is manufactured in Zaire using imported Danfoss compressors and Delco batteries. The refrigerator compartment has 55 litres capacity and the freezer compartment has 20 litres capacity. The refrigerator has been tested and has been demonstrated to be satisfactory. Such local manufacture reduces foreign currency requirements, reduces supply lines and builds up local

COUNTRY	SYSTEM SUPPLIER (REFRIGERATOR MODEL)						POWER SYSTEM		
	ARCO SOLAR (Swafugi)	SOLAVOLT/MARVEL WSR-12-1	BP SOLAR (LEC)	ARG/ELECTROLUX (Electrolux RCW42)	SOLAREX	SOLARFORCE/LEROY SOMER (Frigesol 40)	PV ARRAY PEAK WATTS	BATTERY AMP HOURS	DATES INSTALLED
AFRICA									
Ghana			1				297	450	5/84
Kenya			3				297	450	9/84
									1/85
Tanzania			2				297	450	9/84
									3/85
NEAR EAST									
Yemen Arab Republic						1	160	252	7/84
ASIA				5			230	150	7/84
India									8/84
WEST PACIFIC									
Philippines	1						296	400	11/83
					1		180	660	8/83
Vanuatu		1					144	420	9/83
Solomons		2					144	420	8/83
		1					288	840	2/84

Table 4.3 WHO-EPI Field Trials

knowledge of the technology, which in turn improves operation, maintenance and trouble-shooting.

A complete system with 120 Wp PV array sells for the equivalent of about $4000, inclusive of all taxes and duties.

The users of the PV refrigerators and lighting systems have expressed very favourable views. It has been estimated that in Zaire kerosene refrigerators in practice work on average only 50 per cent of the time and at best 70 per cent in the well-managed health centres. Although initially some problems with the PV systems were reported, PV systems are generally considered by the users to be more reliable than kerosene refrigerators or diesel generators. The FNMA solar refrigerator is shown in Figure 4.3.

South Pacific

The European Development Fund is funding the supply of 23 PV refrigerators for various sites in the South Pacific (10 in Tuvalu, 10 in Papua New Guinea and 3 in the Solomon Islands). The units are being supplied by a Belgian company. The project is being executed by the South Pacific Bureau for Economic Co- operation (SPEC) and installation started in late 1985. It is worth noting that these systems will experience a wide range in climatic conditions. For example, the worst month solar input is only about 2.0 kWh/m^2 per day in the highland areas of PNG compared with over 4 kWh/m^2 per day at the coastal sites (Ref.4.10).

French Polynesia

Since 1978, the French non-profit organization GIE-Soler has been implementing the installation of more than 280 kWp of small PV systems in French Polynesia, including reportedly some 300 refrigerators for medical and residential applications. The project is supported by AFME and the European Commission. It is reported that users are paying from 75 to 80 per cent of system costs and that the project is stimulating further sales (Ref. 4.11).

Sudan
The Sudan Renewable Energy Project (SREP) funded by US-AID and the Special Energy Project funded by GTZ (W Germany) have both provided PV vaccine refrigerators for Sudan. One of the first three units supplied by GTZ failed prior to installation in the field. Laboratory tests conducted by GTZ showed that the vaccine compartment could fall below zero under certain conditions. A further four units (AEG/Electrolux) have now been installed in northern Sudan.

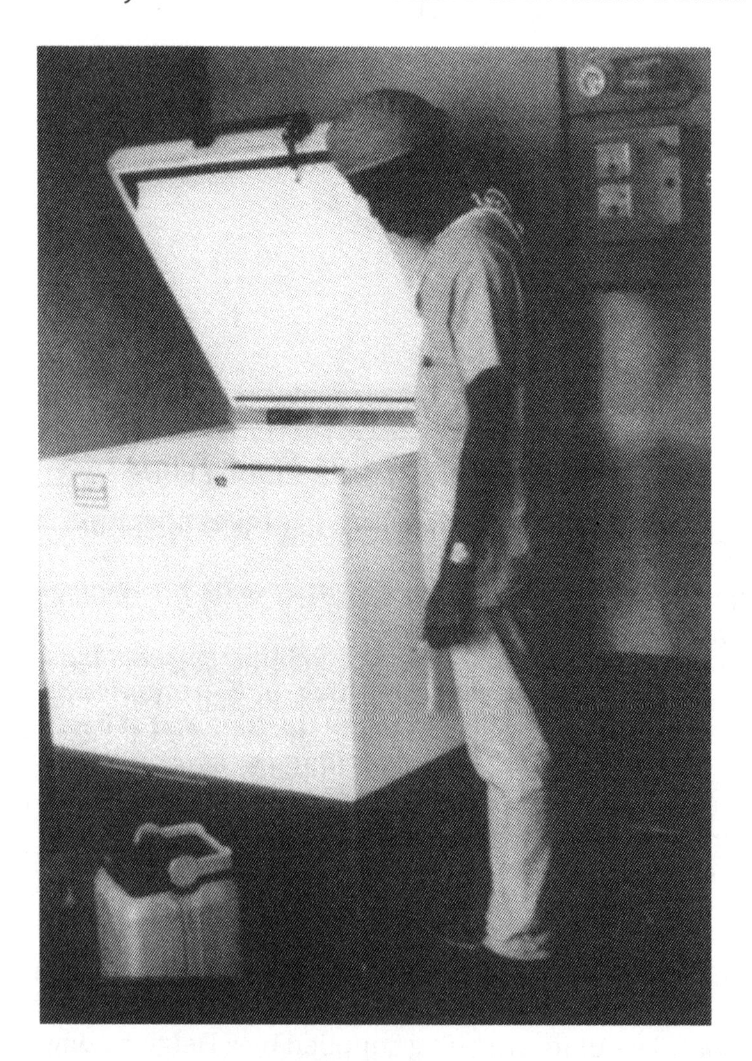

Figure 4.3 The FNMA Solar Refrigerator Manufactured in Zaire

A PV refrigerator supplied by US-AID under the SREP programme is reported to be providing good service at a refugee camp.

Mali
An evaluation has recently been carried out of all known PV refrigerators installed in Mali under various projects (Ref. 4.12). These total 13:3 Arco/WSR, 1 Arco/Polar Products, 1 BP Solar/Lec, 5 Frigesol, 1 Solavolt/Polar Products, 2 SPC/Adler

Barbour. In addition, two Polar Products systems are awaiting installation.

Statistically significant performance data was available for nine installations. Analysis showed that the percentage time the systems were fully operational varied from only 27 per cent (SPC/Adler Barbour) to 79 per cent (Frigesol). Average time between failures for different system types ranged from 4 months to 20 months. These results exclude the recently installed systems. Some of the Frigesol systems and the BP Solar system have been 100 per cent reliable in the short time they have been operating.

Senegal

A 670 Wp medical power system was installed at Mt. Rolland in the Thei district of Senegal in 1982. The PV system provides lighting, refrigeration, improved ventilation with the use of fans and high quality power for laboratory instruments. Overall it has made a significant improvement in health service effectiveness. The dispensary covers a population of 10,000, providing 100-150 consultations per day. The system costs the equivalent of about $20,000, a high price since the system was experimental in nature. The system is also oversized (430 Wp would have been sufficient) due to an overestimate of the load.

Non-Government Organizations

A number of NGOs have installed PV refrigerators in recent years. These include Oxfam (10 FNMA units in Zaire and BP Solar/Lec systems in Uganda) and the International Committee of the Red Cross (ICRC) who have installed several systems in refugee camps. The Save the Children Fund purchased a BP Solar/Lec system which was tested at LESO in Mali before being installed in Douentza. Large orders from German and British based NGOs for solar refrigerators have recently been announced.

4.4 Conclusions

Technical aspects

Although some 800 to 1000 PV refrigerators have been installed to date, experience has shown that the technology has only recently matured. Laboratory testing by WHO and NASA-LeRC has prevented totally inadequate equipment being sent into the field, but some reliability problems are still being experienced. In particular, the sizing of the PV arrays and/or the batteries have been found to be inadequate for actual conditions, in particular in regions of high ambient temperature and poor insolation levels (eg. Philippines, parts of India). Average availability has been round 80 to 85 per cent for systems installed from 1981 to 1983. A summary of observed temperature control is given in Table 4.4.

Systems that have been installed more recently, particularly those from suppliers with previous experience, are being found to be more reliable. Some models are now showing 90 to 100 per cent in-service time, with certain installations operating with 100 per cent reliability for more than two years. Some suppliers have withdrawn from the market (eg. SPC, WSR and Adler Barbour).

The problems of system sizing and load prediction remains a cause for concern. A recent evaluation of tenders for the supply of 23 PV refrigerators for islands in the South Pacific demonstrated that some tenderers proposed PV array sizes and/or battery capacities that would be grossly inadequate. Fortunately, many of the systems tendered were correctly designed. An easily applied method for prospective purchasers to check system sizing would be a significant help.

The ice-making capability of most PV refrigerators commercially available is less than 2 kg/day. Some users have expressed the opinion that this is inadequate for many vaccine cold chains. Battery maintenance has been a common problem with many systems. The possibility of developing a battery-less refrigerator making use of soft-start compressor motors and thermal storage instead of electrical storage should be given more attention as a development project.

The field trials have highlighted the need for a number of relatively minor improvements. These include the provision of door locks on some models and the relocation and/or redesign of some thermostats to reduce the possibility of unnecessary adjustments.

Economic aspects

Very little work has been undertaken on assessing the financial benefits of solar refrigerators using actual field data, but it is important to ensure that investment in PV refrigerators constitutes a sound use of development funds. The WHO-EPI is not an 'economic activity' in the normal sense and so it is not possible to carry out a cost-benefit analysis. The only meaningful analysis involved the comparison of costs of the various options and their likely influence on the achievement of the immunization programme objectives. It is important to note in this regard that the

SYSTEM	Per cent times within temperature bands			No. Days Data
	Correct, 0° to 8°C	High, >8°C	Low, <0°	
1	89.92	10.08	0.00	295
2	86.22	0.00	13.50	711
3	71.72	0.17	28.11	471
4	82.25	11.45	6.29	3057
5	79.11	7.03	13.86	462
6	73.86	0.00	1.59	440
7	62.56	1.37	36.07	219

Source: WHO.

Table 4.4 Observed Temperature Control during WHO Field Trials on Seven Systems (to June 1985)

fixed costs for any immunization programme are generally large compared with the direct costs of vaccine refrigeration.

Kerosene refrigerators used in the vaccine cold chain have an initial capital cost of only $300 to $800, considerably less than for PV refrigerators. With transportation and installation, this may rise to $1500 installed compared with about $6000 for an installed PV system. The operation and maintenance costs of kerosene refrigerators are high however and their reliability is low, sometimes resulting in an availability of only about 50 per cent.

The results of a comparative cost analysis relating to an actual immunization programme in The Gambia are presented in Reference 4.6. based on data collected in 1984 and 1985.

It was concluded that the overhead cost per dose is reduced by $0.06 to $0.07 by using a PV refrigerator, due to the greater reliability. The refrigerator cost per dose is small compared with the overhead cost and is not significantly different between kerosene and PV. The overall cost per dose is cheaper for the PV refrigerator even where the PV capital cost is high.

It is important to note that periods when vaccinations cannot take place result in incompleted and hence ineffective courses of vaccinations. This effect is difficult to quantify but clearly favours the refrigerator with the higher reliability.

A life cycle costing comparative analysis is given in Reference 4.4 which also concluded that the poor reliability of kerosene refrigerators makes photovoltaic refrigeration more economic in comparison. The results of this analysis are presented in Figure 4.4. The data assumed is given in Table 4.5.

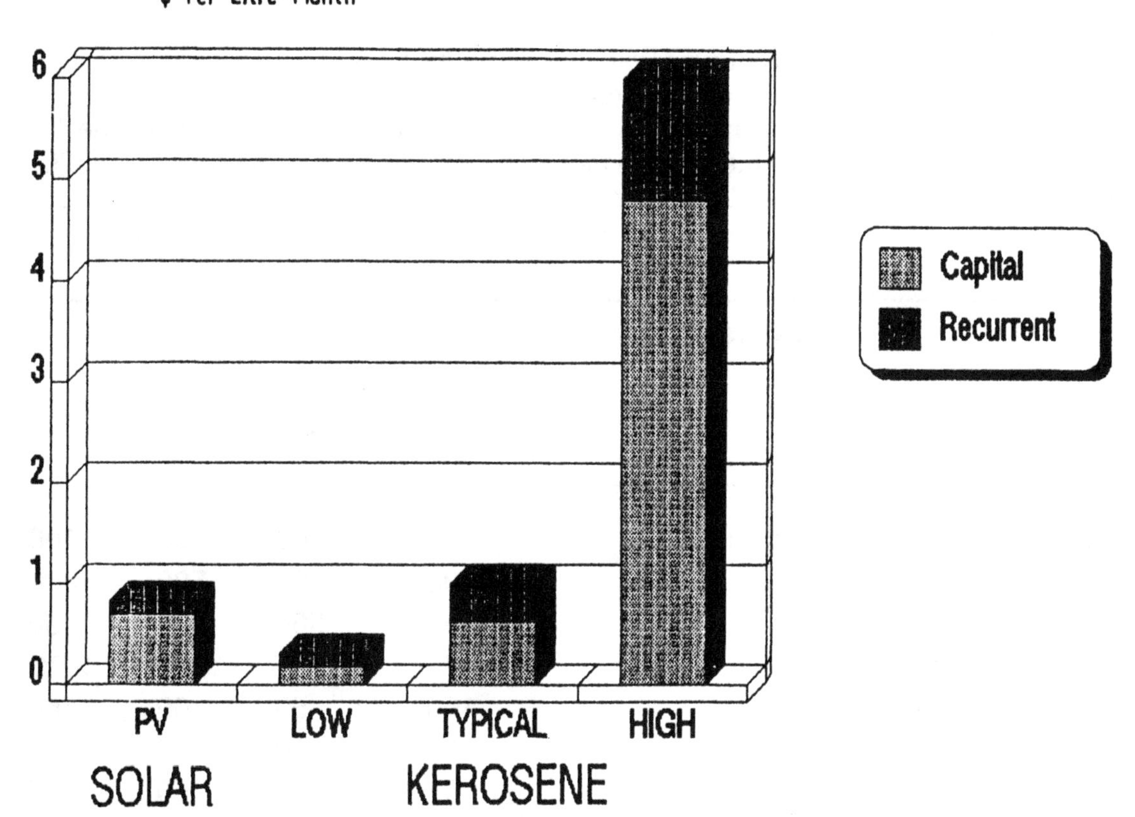

Figure 4.4 Unit Cost Comparison for Kerosene and Solar Powered Refrigerators

Social and institutional aspects

Based on the reported experience of PV refrigeration projects to date, there is no doubt that the systems are widely acceptable to the users. The main need is to ensure a considered approach is taken to project implementation, in particular with respect to:

— project design (selection of systems and sites)
— selection of local implementing agencies
— user training (operation and maintenance and trouble shooting)
— technical support centres serving each region using PV refrigerators
— communications with technical support centre.

The recommendations of the WHO-EPI with regard to procurement of approved equipment and the contractual arrangements for warranties and training of operating and maintenance staff should be followed by all authorities wishing to install PV refrigerators for the vaccine cold chain.

WHO-EPI have now commissioned the preparation of installation, user and repair technician handbooks for photovoltaic refrigerators for use in the vaccine cold chain. In addition regional technician training courses are planed by WHO-EPI. Both of these initiatives should assist with the successful introduction of solar refrigeration into vaccine cold chains.

There is a continuing need to gather data on system performance and therefore efforts should be made to provide the necessary instrumentation and organization required to monitor the systems in the field. The information, both quantitive and qualitative, should be passed on to the EPI co-ordination office in Geneva. The following minimum information should be recorded on a daily basis:

— maximum and minimum internal temperatures of the refrigerator and freezer
— ambient air temperature
— solar irradiation
— system use (kg of ice removed, vaccine removed)
— details of breakdowns or component faults.

For system optimization studies, the electrical energy delivered by the PV array and the energy consumed by the motor/compressor unit is also required. The preferred method of data collection is to use data loggers supplemented by a log book or pro-forma sheets for noting details of system use and reliability.

Refrigerator type:	Photovoltaic	Kerosene		
Parameter		Low case	Typical	High case
Net vaccine capacity (litres)	100	100	100	100
Initial Capital Cost ($)	4500	300	500	1000
Cif and installation ($)	1500	500	800	1000
Fuel costs ($/day)	—	0.25	0.70	3.00
Maintenance costs ($/year)	150	50	100	150
Life time (years)	15	10	5	2
Availability (% time in service)	95	80	50	20

Table 4.5 Data Assumed in Unit Cost Comparison Presented in Figure 4.4

Chapter 4 — References

4.1 'Solar Refrigerators for Vaccine Storage and Ice Making', World Health Organisation, EPI/CCIS/81.5, 1981.

4.2 'Purchasing Guidelines and Product Information Sheets', World Health Organisation, EPI/CCIS/85.4, 1985.

4.3 'The Cold Chain Product Information Sheets', World Health Organisation, SUPDIR 55 AIT 5, 1985.

4.4 A. Derrick and J.M. Durand, 'Photovoltaic Refrigerators for Rural Health Care — Experiences and Conclusions.' *Proc. of the UK-ISES Conference Solar Energy for Developing Countries — Power for Villages*, London, May 1986.

4.5 A F Ratajczak, 'Photovoltaic-powered Vaccine Refrigerator/Freezer Systems; Field Test Results', NASA-LeRC (1985).

4.6 'Pre-Investment Report on Solar Photovoltaic Applications in the Health and Telecommunication Sectors, The Gambia', UNDP/World Bank, March 1985.

4.7 'Solar-Powered Refrigerators for Vaccine Storage and Icepack Freezing: Status Summary June 1985', World Health Organisation, EPI/CCIS/85.4, 1985.

4.8 IT Power Inc., 'Photovoltaic Powered Refrigerators', evaluation report for Meridian Corporation, ref 85153/04, February 1986.

4.9 B McNelis and J M Durand, 'Photovoltaic Refrigerators and Lighting System for Zaire', *Proc. of 6th EC Photovoltaic Solar Energy Conference*, London, April 1985.

4.10 IT Power Ltd. 'Report on the Evaluation of Tenders for Solar-Powered Refrigerators', for South Pacific Bureau for Economic Co-operation, April 1985.

4.11 P Jourde, 'French Polynesia Already in Solar Age', *Proc. of 6th EC Photovoltaic Solar Energy Conference*, London, April, 1985.

4.12 T.J. Hart, 'Technical Assessment of Solar Refrigerators in Mali', Draft Report, January 1986.

5. LIGHTING

5.1 Introduction

Alternative lighting techniques

Lighting is a steadily growing need in the rural areas of developing countries, not only because the population is increasing but also because more people want to be active in the evening. School children need to study and there are new work and leasure opportunities for adults. An important need is for lighting for small commercial enterprises in the streets, such as food stalls, shops and recreational activities. In addition to these residential and commercial needs, there is an associated need for lighting for streets and public open spaces.

In areas where there is no electricity supply, lighting for domestic and commercial applications is usually provided by kerosene lamps or candles. In general, lighting from these sources is of poor quality, expensive and a fire hazard. The best light using kerosene comes from a pressure device (Coleman type), but these are expensive. The more common wick devices (hurricane lamps) produce less than 15 per cent of the light of a 20W fluorescent tube. Due to the high price of kerosene in remote areas, a household may have to spend the equivalent of over $200 a year to operate two kerosene lamps.

Photovoltaic lighting systems would be an attractive alternative to kerosene lamps and candles throughout the areas where it is likely to be many years before regular electricity supplies become available. The key considerations are comparative quality, reliability and cost.

Technical requirements

PV lighting systems have become readily available over the last five years, with manufacturers offering two basic types of unit, one for area lighting, the other for domestic applications.

Area lighting units may be used for street lighting, public open spaces and security lighting. These systems consist of PV array; battery; simple voltage regulator; timing and/or photosensitive switch controls; and an efficient fluorescent or low-pressure sodium or mercury vapour lamp. Several manufacturers offer complete self-contained units including poles with mountings for the lamp and the PV modules and a weather-proof container for the battery and controls. As the poles represent a significant proportion of the total cost, some manufacturers supply only the PV array, lamp, battery and controls, to allow the purchaser to provide the pole from local sources.

Domestic lighting units typically require only one or two PV modules for charging a battery which supplies from one to four fluorescent tubes, from 20W to 40W depending on the application. Some systems are portable, with a lantern unit incorporating a rechargeable battery. Larger systems can be obtained, capable of supplying other end uses such as refrigerators, radios and televisions, but it is more appropriate to consider these systems in the next chapter under the general heading of rural electrification.

Fluorescent lamps are commonly used for both area lighting and domestic lighting systems. Fluorescent (or gas vapour) lamps offer high efficiency, long life and high reliability. They require a 'ballast' and a 'starter' which give a high frequency impulse for starting, followed by much lower power and frequency for normal running. Standard AC fluorescent units may be converted for DC powering (and therefore suitable for PV systems) by changing the ballast and starter components, a relatively simple task.

Field experience

Several thousand PV lighting systems are in use in developing countries. Experience is particularly extensive throughout the South Pacific and more specifically in Papua New Guinea, Fiji and French Polynesia.

There are privately-funded schemes in some countries to enable the benefits of PV lighting systems to be accessible to relatively poor people. In the Dominican Republic, for example, a US-based organization distributes PV lighting systems to villagers in the northern part of the country, with loan finance repaid over two to five years.

5.2 Principal case study — South Pacific

Papua New Guinea

A number of PV systems have been installed in Papua New Guinea (PNG) for communications, lighting, water pumping and medical refrigeration. The total installed capacity in 1982 was about 50 kWp, of which over half was for telecommunications systems (Ref. 5.1). The potential for domestic lighting in the villages over the next 10 years was estimated at about 500,000 single module units of 35 Wp each, a total of 17.5 MWp.

Field experience with the early types of lighting systems (prior to 1980) was not satisfactory, as the battery charge controllers were found to be complex and unreliable. Since then, fully tropicalized charge controllers have been supplied and these have proved reliable. In terms of light quality, a 20W fluorescent lamp was found to give an illumination of 100 lux at a distance 1m below the lamp, whereas a kerosene pressure lamp was found to give only 12 lux at a similar distance. There is general agreement that PV lighting systems are now technically superior to alternative forms of lighting in rural areas. The PNG government has now decided, on grounds of safety and reliability, to use PV lighting systems for official patrol posts in the villages.

Regarding economics, reference 5.2 presents the results of a survey which was conducted among 30 village houses to assess the cost of kerosene-fuelled lighting systems for comparison with PV systems. A typical household was found to use one hurricane lamp and one pressurised lamp. Annual costs were found to be the equivalent of about $230 (1981 prices). The Present Worth at 10 per cent discount rate of five years expenditure is thus about $970. The comparison PV system, comprising a 35 Wp module, battery, regulator and two 20W fluorescent lamps, could be installed in 1981 for about $775, with the module guaranteed for five years and the battery for two years.

Although the PV system was cheaper than the kerosene lamps over a period of five years, it was recognised that the capital cost would be a formidable barrier to the potential user. In view of the improved technical quality of the lighting and the benefits to the national economy, the study concluded that lending institutions in PNG should be encouraged to provide finance to customers wishing to purchase PV lighting systems.

Fiji

Most Fijian villages, consisting typically of 10 to 50 homes, are located at too great a distance from the exsisting national grid to be connected at acceptable cost. For many years, the Fiji Public Works Department has been installing small diesel generating plants (7.5 to 15 kVA) in villages to provide electricity mainly for lighting purposes. To date, however, less than 100 villages out of more than 2000 in Fiji have received such installations, due to limited technical and financial resources.

In view of the potential advantages of PV lighting systems, a programme was initiated in 1982 by the Fiji Department of Energy to test and demonstrate PV domestic lighting systems (Ref. 5.3). A specification was drawn up for a system comprising two 15W fluorescent lamps, a 35 Wp PV module, a battery charging unit for four Ni-Cad 'D' cells for powering a portable light or radio and a gel-type main battery. The system is a total package including all components and instructions for field installation and use.

After international tendering procedures, a total of 100 systems were ordered from South Pacific Solar (USA). These were delivered in early 1983 and installed in three villages. Although each system was partially subsidized, all users had to make a contribution to capital costs and pay a regular sum for maintenance and improvements. The systems have been widely welcomed by the users.

An economic study was also carried out in 1983 comparing the costs of kerosene lighting with diesel generator-based systems and PV systems. Fuel costs for wick and pressure lamps (kerosene or benzine) for a typical household were found to range from the equivalent of $120 per year to over $180 per year, depending on the remoteness of the site and the associated fuel transport costs. Fuel costs for urban users were considerably less, typically $90 per year. A life of five to seven years is reasonable for pressure lamps with a replacement cost of about $40 each. Wick lamps have an initial cost of from $5 to $15 and have an indefinite life unless broken.

Diesel generating sets were found to cost typically $2000 per year for fuel and maintenance in a village of 30 houses. The initial cost was $550. Installation would cost a further $50. Assuming a battery life of four years, the replacement costs over 10 years would be $220. Annual maintenance was assumed to cost about $15 per system, to cover an annual visit by a qualified technician to a group of 30 systems.

Based on the above costing assumptions, the Present Worth of all costs over a 10 year period for providing lights for a village of 30 homes for the three methods is as follows:

— Kerosene and benzine lamps	$28,000
— Diesel-generator system	$26,000
—PV systems	$25,000

The study concluded that for remote villages where lighting cannot be provided by grid-delivered electricity, photovoltaics are marginally better economically and should be examined carefully in the light of their advantages over diesel electrification and kerosene or benzine lamps.

Initial results from the demonstration projects in Fiji have been favourable, with a high level of user acceptance and a reasonable prospect for economic success.

French Polynesia

In 1983, the French Polynesian firm of SOL E.R. received a rush order to supply electric power systems to 200 homes on the remote island of Faaite. Within a week, this locally-run renewable energy firm had assembled and despatched by boat 200 PV modules, 40 regulators, 60 batteries, 350 fluorescent lamps, 5 refrigerators, 5 washing machines, 3 water pumps, and other electrical equipment. This equipment was installed over the next four weeks (Ref. 5.4).

This report of the rapid electrification of an island is apparently typical of many similar projects undertaken in the South Pacific. Most of the equipment, apart from the solar cells which are imported from France, is designed and built in Tahiti. The major electrification programme for the remote islands started in 1982, after a three-year assessment and testing phase. By mid-1983, over 550 houses had been equipped with PV systems, usually with from 2 to 20 modules mounted on the roof, at a cost ranging from $700 to $7000 (average $1500) per house. Most of the houses thus equipped were traditional Polynesian thatched-roof homes.

The purchasers of these PV systems receive a 25 per cent capital grant from the government. The balance is raised by the householders themselves, either from their own resources or with the help of a commerical loan. The systems are reported to be cost-effective in comparison with diesel generators, which are particularly expensive to install, operate and maintain on remote islands.

An important feature of the PV electrification programme in French Polynesia has been the training of technicians in system design, construction, installation and maintenance. In addition to many demonstration projects, a comprehensive programme has also been undertaken by the government to inform the public in general of the potential offered by PV and other renewable energy systems (Ref. 5.5).

5.3 Other lighting projects

Mali

A classroom in Mali was equipped with a PV-powered lighting system in 1980 to enable it to be used for evening classes. Although five years later the system is reported to have operated well, requiring little maintenance, the high initial cost was seen as a major barrier to widespread replication, in view of the limited funds available for the development of rural education.

United Arab Emirates

PV-powered street lights have been installed in a number of countries in the Middle East, including a project in Dubai involving 21 street lights and a high-mast flood light for a traffic circle. The equipment was supplied by Mobil Solar Energy Corporation (USA) in 1983. Each street light consists of a 20W fluorescent tube, two 35 Wp PV modules, battery with regulator and controls. The high-mast system consists of eight 400W high-pressure sodium vapour (HPS) lamps powered by a 15 kWP PV array and battery bank, with inverter and control system.

Five commercially available fluorescent tubes were tested at the design stage and large differences in illumination efficiency were observed. The most efficient tube and ballast combination was chosen for the actual installation. The street lights give a good level of illumination and are reported to be operating reliably.

There were initial problems with the HPS lamp due to the inverter used which was overcome by replacing it with one of a different type.

China

Three hundred and fifty million people in China live in villages without an electricity supply. Many are nomadic herdsmen living in Inner Mongolia and other remote provinces using candles, kerosene and the oil derived from sheep for lighting. The costs of transporting kerosene are very high and extending the grid would be out of the question in many cases. PV lighting systems however offer particular advantages and experience of recent years indicates that such systems would be technically and economically feasible. In 1985, it was reported that there were over 2000 PV lighting systems of various types being used in China (Ref. 5.6).

Dominican Republic

The US-based organization Enersol Associates Inc. has been helping villagers in the remote parts of the Dominican Republic install PV lighting systems since the first was installed in April 1984. A total of 15 systems had been installed by the end of 1985 and a rural solar co-operative formed to facilitate financing and education through information exchange. A hardware store and service centre has been opened to provide a continuing back-up service to users. Peace Corps personnel have helped run training workshops (Ref. 5.7).

With assistance from US-AID and other sources, Enersol Associates have established a revolving fund to finance the co-operative's activities. Local users pay for their PV systems over a period of two to five years with reasonable interest rates.

The calculated cost of electricity provided by the alternative technologies available for remote areas are as follows: (Ref. 5.5).

— PV system	$1.00/kWh
— Car battery	$2.50/kWh excluding costs of transporting the battery from the nearest place where it can be recharged.
— Kerosene wick lamp	$3.50/kWh
— Kerosene pressure lamp	$14.00/kWh
— Dry cell battery	$30-$35/kWh

Zaire

An area where lighting can make a substantial impact is in rural health clinics and medical centres. In Zaire the Département de la Santé Publique is installing 850 lighting systems as well as 100 photovoltaic vaccine refrigerators. The project is funded by the European Development Fund.

5.4 Conclusions

Technical aspects

PV lighting systems covering a wide range of sizes and types are widely available as standard products. The components required for a typical domestic lighting system are listed in Table 5.1. Such a system would provide up to 200 Wh/day of useful energy for lighting given a solar input of 6.0 kWh/m^2 per day. The 20W lamp could be used for 3 to 5 hours every night for general activities and the 7W lamp could be used for 8 to 12 hours for security. The battery provides about three days storage.

Many of the smaller systems for domestic use are portable, which makes them particularly suitable for use in place of kerosene lamps. The introduction of long-life rechargeable Ni-Cad batteries is an interesting development in this regard.

The battery charge controllers used for some early designs of PV lighting systems were found to be unreliable, but now fully tropicalized units are supplied which have proved very reliable in practice. The reliability and efficiency of the ballasts used in commercial fluorescent lamp units have been found to be variable. Careful selection of this component is therefore essential, particularly as it accounts for up to 75 per cent of the cost of the lamp. The lifetime of the DC ballasts used for some 60 DC fluorescent fixtures tested as part of the NASA-LeRC PV medical refrigerator programme was found to be less than five years. Further experience of the lifetime of these components under field conditions is needed.

Specification	Life (years)	Price ($)
1 PV module — 40 Wp	15	250-300
Battery — 12V/105Ah	4	50-80
Fluorescent lamps	2	50-70
Battery charging controller	5	50-200
	Total	400-500

Table 5.1 Typical PV Lighting System for Domestic Use

PV-powered street lights and security lights are also available from several manufacturers. These units generally use low-pressure sodium vapour or high-pressure mercury vapour lamps. Some problems connected with the need to adapt standard AC units for DC operation have been experienced and there is need for further development of suitable ballasts.

Economic aspects

Lighting is not a directly economic activity and therefore a cost/benefit analysis is not posssible for this application. A number of cost comparisons for alternative lighting methods, including the projects detailed earlier, indicate that PV lighting systems offer the cheapest solution for lighting in villages where grid electricity is not available.

It should be noted that the light output of a kerosene pressure lamp is about 200 lumens. That for a kerosene hurricane lamp is about 80-100 lumens, whereas the light from a 20W fluorescent tube is about 1000 lumens. Also the life of the PV modules (the most expensive part of the PV system) is at least 15 years, much longer than the life of kerosene lamps.

Provided suitable means are available to finance the initial cost with repayments over say 5 years, PV systems should be widely attractive on both cost and performance grounds to potential users in the villages.

Social and institutional aspects

For the successful introduction of PV lighting sytems, the potential users need first to be convinced of their technical performance and reliability. This requires demonstration systems to be available, possibly as mobile units to be taken from district to district.

There are then two major institutional requirements:

a) The provision of finance, probably through the provision of a subsidy and low-cost loan, repayable over two to five years;
b) The provision of technical support, in particular for the supply of spare fluorescent tubes and ballast and batteries.

In many cases, it would be preferable for the implementing organization to establish local co-operatives, who can arrange for the administration of funds and the provision of technical supports. Training for key personnel would be needed. The local co-operative would need to be able to refer major problems to a central resource centre.

Although in the short and medium terms most countries would need to import the PV modules, most other components could be locally manufactured and assembled, thereby greatly reducing the foreign exchange requirements whilst at the same time building up local technical skills.

Chapter 5 — References

5.1 G.H. Kinnell, 'Solar Photovoltaic Systems in the Development of Papua New Guinea', *Proc. of the Fourth EC Photovoltaic Solar Energy Conference*, Stresa, May 1982.

5.2 K. Maleva, 'Feasibility Assessement for Photovolaic Cells Replacing Kerosene Lighting in Papuan Villages', Report No. 7/81, Energy Planning Unit, Department of Minerals and Energy, Konedobu, Papua New Guinea.

5.3 H.A. Wade, 'The Use of Photovoltaic Systems for Rural Lighting — an Economic Analysis of the Alternatives', *Proc. of the Solar World Congress*, Perth, August 1983.

5.4 D.O. Hall, '*Electrifying an Island a Month*', Earthscan Feature, London, 1984.

5.5 'Renewable Energy in French Polynesia', by SOL E.R. and CEA-Ger, Pappeete, French Polynesia, 1985.

5.6 Zhu Gangi, 'Recent Experience on Solar Energy Utilisation in China', *Proc. of UK-ISES Conference C42, Energy for Development — Where are the Solutions?* Reading, UK. Dec 1985.

5.7 'Newsletter No. 4', Enersol Associates Inc. Somerville, USA, Jan 1986.

6. RURAL ELECTRIFICATION

6.1 Introduction

Conventional approaches

There are two main techniques used for rural electrification at present: either the main electricity grid is extended to cover the selected area, or a diesel generating station is established to serve a small network not connected to the grid. Both approaches have associated economic and technical problems. Extending the grid over long distances is expensive and, in the initial years at least, the loads are often small, resulting in load factors and stability problems for the system as a whole. Maintaining long transmission lines over difficult terrain presents difficulties for the utility.

Diesel generators require regular supplies of fuel, which often presents problems for remote areas, particularly at certain seasons of the year when roads are practically impassable. Moreover, any national fuel shortages are likely to impinge more severely on the rural areas. In addition to the provision of fuel, it is often the case that the operating utility finds difficulty in retaining competent operating and maintenance staff, since as a parastatal organization they are unable to offer competitive salaries. Even given the necessary staff, it may be practically impossible to obtain the necessary spare parts to keep the engines running.

Rural electrification is regarded as a development priority in most developing countries, for social and economic reasons. Although vast sums are expended each year on rural electrification projects, it will nevertheless be many years before villages that are a long way from the main electricity grid lines or from the nearest all-weather road will benefit from a reasonably reliable and affordable electricity supply.

There is no doubt that new and renewable energy sources would be preferred for rural electrification, if systems were available at acceptable costs and with proven reliability. In some locations, wind generators or micro-hydro systems may be feasible, but in general photovoltaic systems would be favoured if the costs were right, since they involve no mechanical moving parts and require only simple maintenance.

Design and load estimation

For many households in developing countries, the main use for electricity would be for lighting. Many thousands of small PV systems dedicated for lighting applications are already in use worldwide, as discussed in the previous chapter. For many households, however, electricity would be useful for many other applications. Some of these could also be covered by small dedicated stand-alone systems, such as sytems for water pumping or rice milling as discussed elsewhere in this survey, but in general it would be preferable for rural electrification schemes to be non-specific, with the electrical power available for any small-scale application.

The great majority of users would be domestic households. Based on data presented in References 6.1 and 6.2, for design purposes the peak demand and daily energy consumption for a typical household would be as shown in Table 6.1. Although in general it may take several years to build up to the power and energy levels shown, as most households could not afford all the appliances involved at once, it is prudent to plan a rural electrification scheme for the situation likely to arise within a few years.

There will of course be a number of connections which exceed this load estimate, as well as many connections which are much less. There are differences between countries, depending on standard of living and social patterns (eg. number of persons living in one household). In addition to estimating domestic load, it is necessary to calculate specific loads for commercial, institutional and industrial applications.

Household versus central systems

To date, most PV demonstration projects for rural electrification have consisted of either small packaged systems for lighting or other specific end uses, or larger centralized projects serving a whole village. For a demonstration project, it is administratively easier to install a single central unit to serve a whole village. There is no doubt that the larger the system, the greater the publicity. There are however a number of important advantages to be gained by installing individual household systems rather than a central PV plant. The main reasons for preferring household systems (from Reference 6.1) are as follows:

(a) The PV arrays can be roof-mounted, out of reach of livestock and people and not taking up valuable land;
(b) A distribution system is not required to take the power to each house, an expensive item if the houses are widely spaced;
(c) No metering system is needed, thereby avoiding the associated administrative costs for meter readers and billing computations;
(d) With no distribution system, the problems associated with unauthorized connections and theft of electricity are avoided;
(e) A centralized system may soon prove unreliable, due to its complexity and to possible overloading aggravated by unauthorized connections; failure of the system affects everyone, whereas failure of a household system affects only one consumer.

Although a centralized system may appear a little cheaper in initial cost, even after allowing for the cost of the distribution system, this is heavily outweighed by the reduced operational problems and administrative costs associated with separate household systems.

System features

If PV generators are used for rural electrification, whether through household systems or centralized plants, the DC output would normally need to be converted to AC at the national standard voltage and frequency. Although most of the loads such as lighting and domestic appliances could be designed or adapted for DC operation, this would be unpopular with consumers, who would wish to have the freedom to buy cheaper standard AC products from the normal suppliers in the markets. DC equipment would introduce too many complications. A typical PV system for rural electrification by means of a central plant is shown schematically in Figure 6.1. The schematic arrangement for a household system is similar, as shown in Figure 6.2.

Central plants would need to have battery storage for about three to five days supply, depending on the Loss of Load Probability (LOLP) level considered to be

Load	Power (W)	Duration (hrs/day)	Energy (Wh/day)
Lights 3 × 20 =	60	5	300
Fans 2 × 60 =	120	6	720
TV	60	4	240
Iron	200	1	200
Miscellaneous	60	1.5	90
Total	500		1550

Table 6.1 Typical Design Electrical Load for Rural Household

appropriate. The battery storage for household system could perhaps be less, say two to three days supply, since energy management procedures could more readily be introduced by the user when an alarm on the system indicates that battery charge is low.

An energy meter would not be required for a household system, since it would be most appropriate to charge the consumer a constant monthly or quarterly rent for the system. Although the consumer would need to understand the load limitations of his system, there would be no particular advantage in trying to economise on the use of electricity, at least all the while the battery state-of-charge was within its normal operating range.

Commercial equipment

There are no standard systems for central PV plants, although a number of demonstration plants have been built and all the major PV manufacturers are able to offer a design service for such plants. Several of the PV pilot and demonstration plants sponsored by the European Community are for rural electrification (eg. Aghia Roumeli, Crete, Greece, 50 kWp and Kaw, French Guyana, 35 kWp). In addition to the central plants, there is also an EC-supported demonstration project currently in progress which provides for the installation of small stand-alone PV generators for some 40 houses in the south of France and in Corsica involving three standardized designs: 400 Wp, 800 Wp and 1200 Wp.

A number of village electrification projects involving photovoltaic systems have been undertaken in developing countries. Two villages in Indonesia have been equipped with central PV plants and there are similar schemes in Tunisia, French Guyana and Senegal. There are also a number of villages on Greek islands in the Mediterranean equipped with central generators.

In addition to these and other central plants, the alternative disaggregated approach using stand-alone systems for household and specific end uses has been followed in French Polynesia, Gabon and elsewhere.

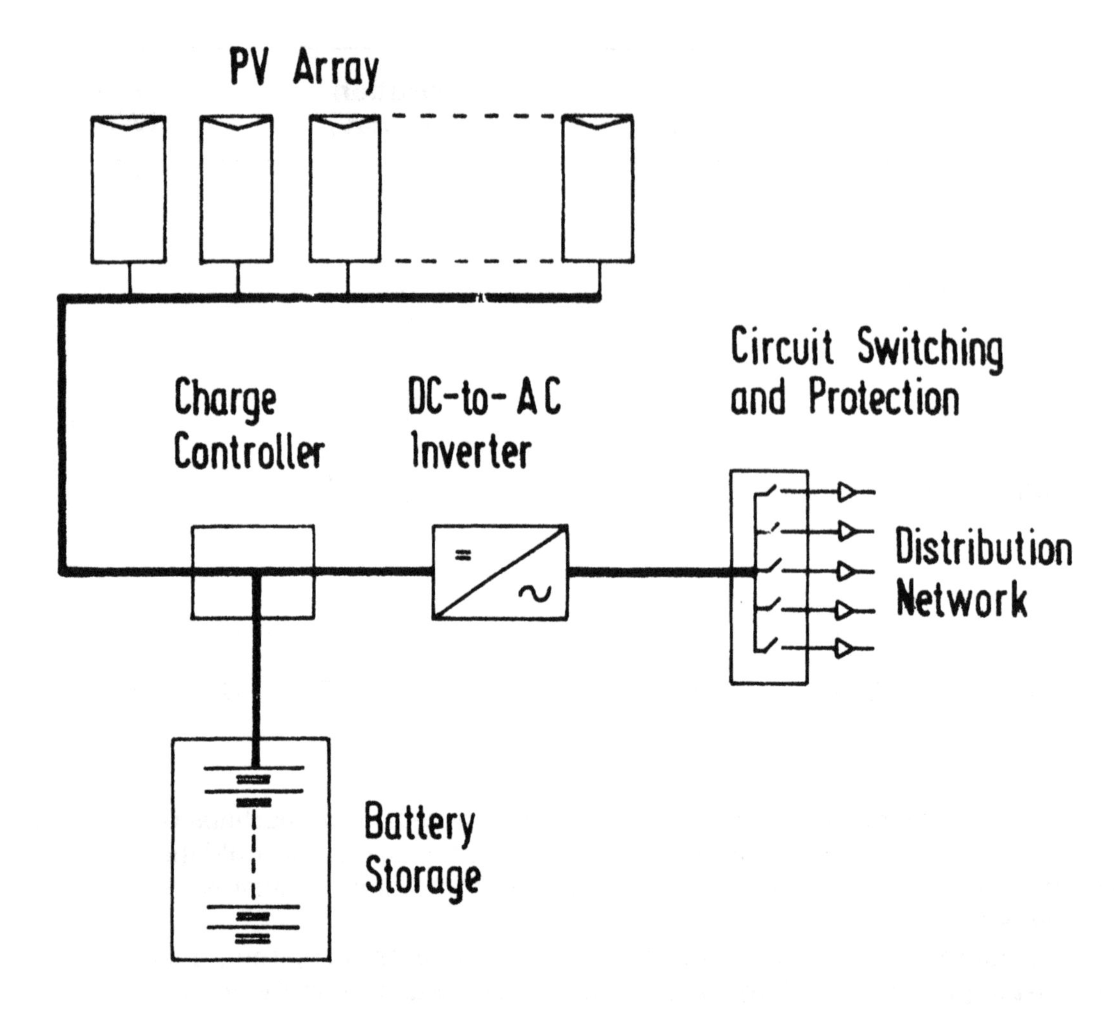

Figure 6.1 Schematic of PV Central Plant for Village Electrification

6.2 Principal case study — Indonesia

Background

A joint Indonesian/West German project called 'Solar Village Indonesia' was initiated in 1979. The project included a number of collaborative activities to develop systems for the direct and indirect use of solar energy in a tropical environment, with the long-term objective of improving living conditions and agricultural output in the rural areas (Ref. 6.3).

Four field test sites were selected for different activities within the overall programme. The village of Picon in West Java has 350 inhabitants (60 families) and was provided with two main systems:

— two fixed-bed gasifiers and a village biogas plant which together produce gas to operate engines which drive electrical generators (60 kW)

— a 5.5 kWp PV generator producing power for 10 irrigation pumps, carpentry workshop, rice processing and village lighting (see Figure 6.3).

The village of Citius on the northern coast of Java has a population of 780 people, the majority fishermen. The ice needed for fish conservation was brought daily from 20 km away. A 25 kWp PV system was installed for the following applications:

— a vapour compression ice-making plant for producing 500 kg of crushed ice per day

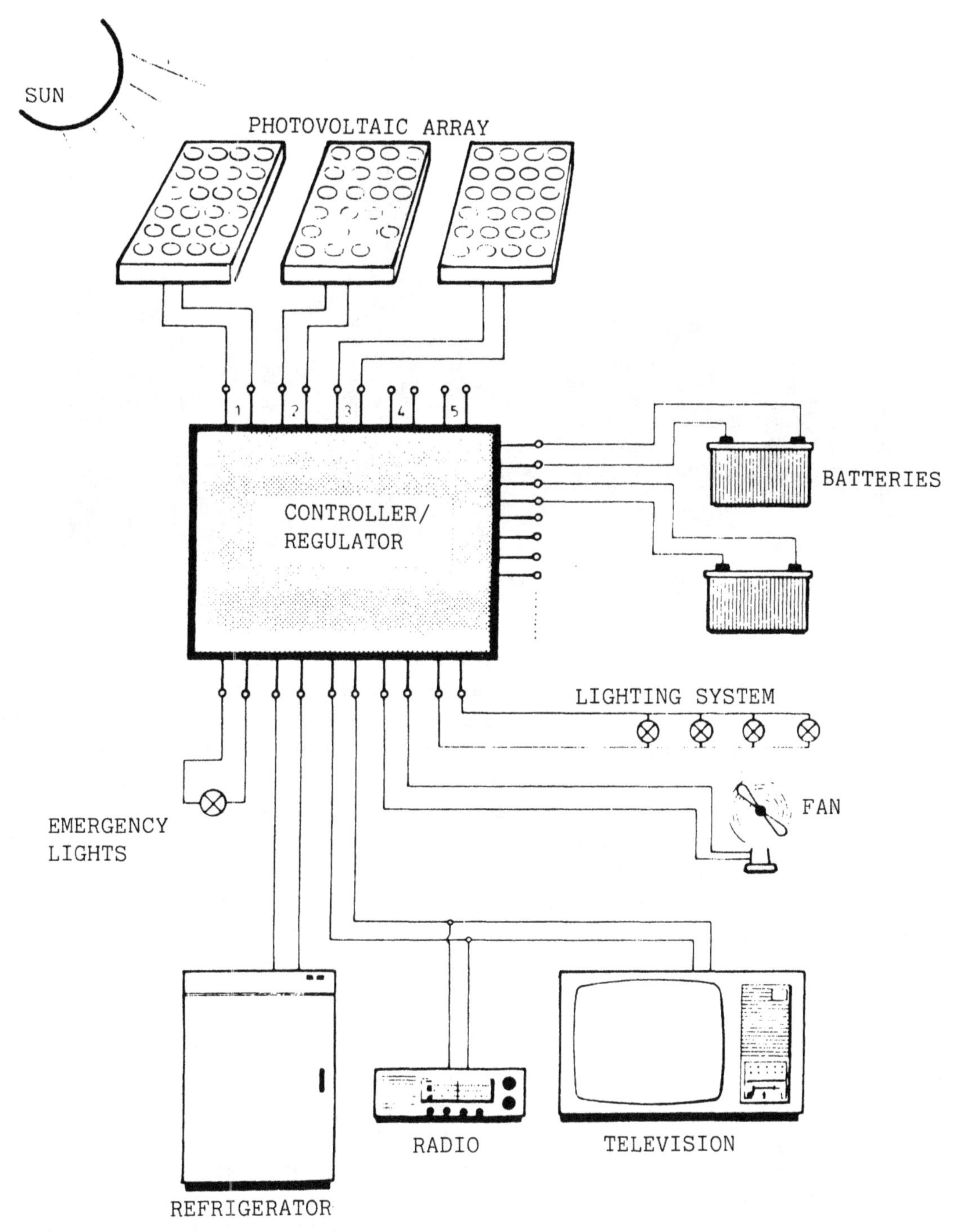

Figure 6.2 Schematic of PV System for Household Electrification (Source: Solar Energie Technik)

— A Reverse Osmosis desalination plant providing 5 m^3 of drinking water per day using brackish water from a well
— a comprehensive data-recording system.

In addition, two PV-powered navigation buoys were installed to mark the harbour entrance and a PV-powered television set provided for use by the villagers.

Puspiptek is an Indonesian government research station occupying a 350 hectare site situated near Jakarta. In Puspiptek, five solar thermal pilot plants are installed for research and development purposes.

Figure 6.3 5.5 kWp PV Array at Picon, Indonesia

The fourth field test centre is on the remote island of Sumba south-east of Java. Many villages on this island do not have any water at all, so that the inhabitants in some cases have to walk up to 15 km to the nearest well. The aim of the activity on Sumba is to show that photovoltaic pumping systems offer a realistic solution to water supply problems encountered in remote, arid areas. Three PV pumping systems have been installed as part of the project: a 5.76 kWp DC system at Gollu Watu pumping water from an underground river 28m deep for a population of 6000 throughout the year; a 3.65 kWp AC system at Pemuda pumping water from a 60m deep well for a population of about 2000 in the dry season; and another 3.65 kWp AC system at Wee Muu providing drinking water from a 37m deep borehole for about 2000 throughout the year.

Experience

The 'Solar Village Indonesia' project (Ref. 6.4) is being carried out in three phases, as follows:

Phase I — Definition phase, completed September 1980

Phase II — Implementation phase, completed 1983

Phase III — Performance testing, operation and maintenance, evaluation of systems, from 1983.

A joint working group has been established to undertake the tasks included in Phase III. The group consists of a test and evaluation team, an operation and maintenance team, experts for special problems and tasks, and an expert from W. Germany permanently on site.

Phase III is still in progress, but it is envisaged that when the evaluation studies have been completed it will be possible to replicate similar systems at other sites in Indonesia. Future systems will incorporate such improvements as have been found desirable and will be manufactured as far as possible in Indonesia.

Preliminary results from the 5.5 kWp PV plant at Picon village have indicated that overall system performance and efficiency has been strongly influenced by the poor partial load performance of the inverter originally installed. A new inverter with improved performance has now been incorporated.

6.3 Other rural electrification projects

Tunisia

A solar village project involving PV, wind and solar heating systems has been operating in Tunisia since 1983 (Ref. 6.4). The main photovoltaic generator consists of a 29 kWp array with battery bank and inverter to deliver power at 220V, 50 Hz, to the village. There is also a 1.4 kWp stand-alone system serving a farm and two 1.4 kWp drip irrigation systems. The solar village was developed with the assistance of the NASA-Lewis Research Center (USA). Operation and evaluation is now the responsibility of the Société Tunisienne de l'Electricité et du Gaz (STEG).

French Guyana

A 35 kWp PV plant to serve the isolated village of Kaw in French Guyana was installed in 1982 as part of the European Community photovoltaic pilot plant programme (Ref. 6.5). The PV system replaced two diesel generator sets which had proved unreliable and expensive to operate. The system consists of 528 PV modules, each rated at 66 Wp, arranged in 44 panels of 12 modules each. The battery bank provides 410 kWh storage capacity to ensure continuous power supply throughout the year to the village. A high efficiency inverter is used, giving three phase and neutral output at 380V, 50 Hz.

The system is operated and maintained by the French electricity utility, EDF. Special training for the EDF technicians responsible was provided by the main contractor, Seri Renault (France) and the inverter supplier, Jeumont Schneider (France). The operation has been practically trouble free since commissioning in January 1983, fulfilling all expectations.

Senegal

A PV/wind hybrid generating system started operating at the village of Naiga Wolof, Senegal in February 1983. The PV portion is rated at 5.5 kWp and the wind generator is rated at 4.5 kW. Initially the system provided power only for public lighting. Then, in April 1984, a water pump was added, followed by two refrigerators and additional lights. The system is eventually intended to supply residential lighting, a communal television, carpentry equipment, a sewing machine, a grain mill and an ice maker.

The PV portion of the system is reported to have operated reliably without the need for a full-time system operator.

French Polynesia

As reported in the previous chapter, there is a major rural electrification programme for remote areas of French Polynesia.

The programme is supported by the French Atomic Energy Commission (CEA), the French Agency for Energy Management (AFME) and the Government of French Polynesia. The majority of the PV systems installed are relatively small, designed to be stand-alone systems for individual households. In addition to the many small systems dedicated for lighting, systems are also being installed to provide for other end uses. A typical household system provides enough power for three 13W fluorescent lights, an 80W television set, a fan and a small refrigerator. The cost of such a system is about $2000, including taxes, but a government subsidy is available covering 50 per cent of the module price. Finance is available for purchasers to enable them to spread the cost over a five year period.

6.4 Conclusions

Technical aspects

There is limited experience available to date on the performance of larger centralized PV plants for village electrification. Some systems have worked well, others have experienced problems with inverters and other components. No standardized designs exist as yet and care must be taken at the design stage to ensure that only components with proven reliability are used in systems that are to be installed at remote sites.

There is however a considerable volume of experience with smaller household systems, with many hundreds of installations in French Polynesia and elsewhere. Standardized designs have been developed and performance is generally reported to be very satisfactory.

Economic aspects

There is little economic data on the cost of installing, operating and maintaining centralized PV plants for village electrification. There is however extensive information on the economics of household systems. The current (1986) installed cost of a small (less than 500 Wp) stand-alone PV generator (including batteries and inverter) is at least $20/Wp if bought as a one-off item. If however standardized systems are bought in large quantities and integrated with the building, it is possible to get installed system costs down to around $15/Wp. Based on the low price scenario for PV modules and system presented by Starr and Hacker in Reference 6.6, a cost projection for standard 360 Wp PV systems suitable for stand-alone household AC generators can be constructed, as shown in Table 6.2 (Ref. 6.2). The price per peak Watt is shown as falling from a current level of $15 to $5 by the year 2000.

Component	1985	1990	1995	2000
Modules (360 Wp)	1800	1080	810	540
Battery (299 Ah/48V)	1600	1200	800	440
Inverter/Regulator	1000	600	300	250
Building-related costs, transport, installation, etc	300	300	300	300
Supplier's mark-up	700	470	330	270
Total installed system price	5400	3650	2540	1800
Price per peak Watt	15	10	7	5

Note: all costs in 1985 US Dollars

Table 6.2 Cost Projection for Household PV Systems (from Reference 6.2)

It is now relatively straightforward to calculate the average unit cost of electricity produced by the PV generator for the range of system costs shown in Table 6.3. Annual maintenance expenses are assumed to be 1.5 per cent of the initial capital cost. Administrative expenses are assumed to be $10 per system per year, given that there are a large number of systems within each administrative area. The average electricity cost is based on a 20-year period of analysis and a 10 per cent discount rate. The average solar input is assumed to be about 5 kWh/m^2 per day. Each system is assumed to give a useful energy output of 540 kWh/year. The resulting average energy costs are as follows:

Installed system cost ($/Wp)	15	10	7	5
Average energy cost ($/kWh)	1.34	0.90	0.64	0.46

These unit energy costs may then be compared with the real (unsubsidised) cost of electricity supplied by alternative means, such as grid extension or by diesel generators. Grid extension costs depend on two main factors; the distance the feeder line has to be extended to serve a new area; and the number of connections served by the new feeder. Based on typical cost data, the unit energy costs for grid extension schemes are as set out in Table 6.3 for three values of feeder line length and five values of the number of connections per feeder.

The unit energy costs for the PV and grid extension approaches are compared in Figure 6.5. Grid extension is the cheaper alternative for the near future except for the occasions when it is necessary to provide electricity to a limited number of connections involving long feeder lines. Indeed, for providing power to a few isolated houses, it is likely that a PV system will be cheaper than the grid if the grid is more than a few hundred metres away.

Diesel generation costs are summarised in Table 6.4 based on data published in Reference 6.4. The unit energy cost for a system supplying 540 kWh/year (about 1.5 kWh/day) ranges from $1.00/kWh for the low cost case to over $2.50/kWh for the high-cost case. Larger diesel generators serving whole villages typically give costs from $0.60/kWh to $1.50/kWh, given reasonable maintenance and no major interruptions in the supply of fuel and spare parts.

From this data, it is clear that PV systems are cost-competitive today with small diesel generator systems. They can also be competitive with larger diesel generator systems in places where fuel supplies and maintenance present major difficulties. PV systems will become increasingly competitive with grid extension schemes as the cost of PV modules and other components continues to fall with improved technology and larger volume production.

	Feeder length km	No. of connections per feeder 100	200	500	1000	2000
Unit energy cost	10	1.00	0.67	0.47	0.40	0.37
in $/kWh	20	1.69	0.97	0.61	0.47	0.40
	30	2.35	1.34	0.74	0.54	0.44

Table 6.3 Unit Energy Cost for Grid Extension (from Reference 6.2)

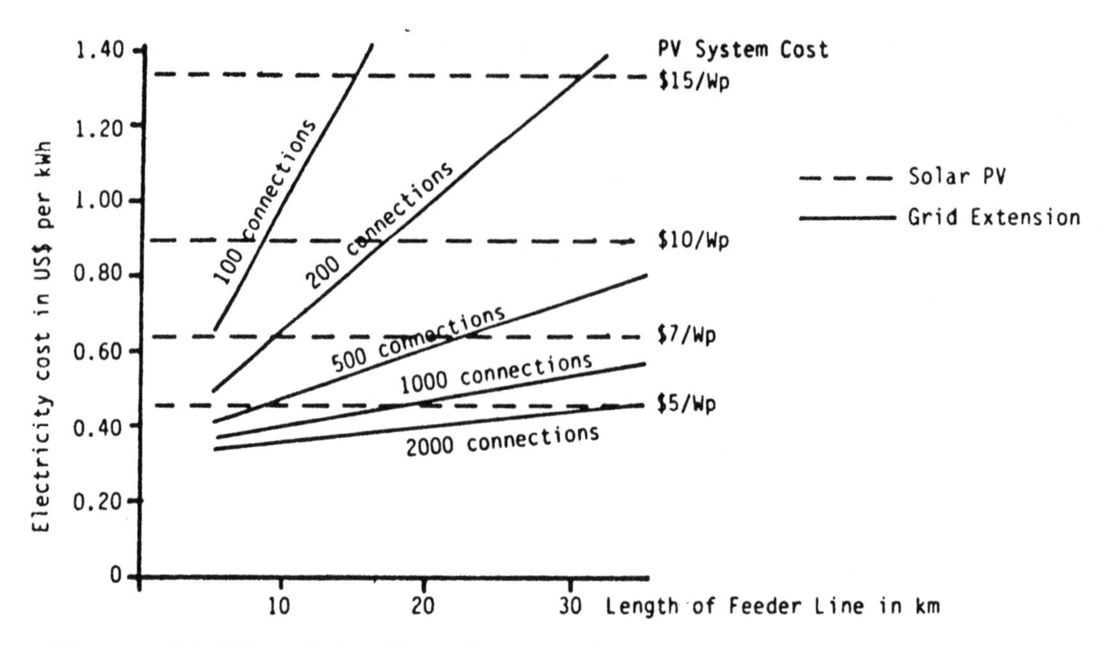

Figure 6.4 Electricity Unit Cost — Grid Extension and PV Systems

		Low cost	High cost
Size of diesel (kW)		2.5	2.5
Life of system (years)		7	7
Capital cost ($)		1000	3000
Fuel cost ($/litre)		0.40	0.80
Overall efficiency (%)		20	15
Operating O & M ($/year)		200	400
Unit energy cost in $/kWh for:			
	0.5 kWh/day	2.30	6.00
	1.0 kWh/day	1.50	3.30
	2.0 kWh/day	0.80	2.10

Table 6.4 Diesel Generator Costs

Social and institutional aspects

Whatever method is adopted for rural electrification, costs are bound to be high. Many developing countries have established rural electrification boards (REBs) to undertake the necessary planning, implementation and operation of rural electrification schemes. The REBs normally require substantial external funding, since it is widely recognised that rural electrification projects can rarely be self-financing. The selling price of electricity to rural consumers has to be kept at a low level, comparable to that obtaining in urban areas, since (a) most rural consumers are very poor and (b) any significant disparities would generate strong political and social pressures. The cash flow for a typical REB with an expanding programme is thus bound to be poor, particularly as it often takes several years for the loads and associated revenues to build up.

Rural electrification schemes therefore have to be economically appraised and justified on broader grounds than simply costs and revenue from the electricity system alone. The value of the social benefits that accrue to the community as a whole, through the raising of living standards, improvements to land and labour productivity and the generation of new employment opportunities, all need to be assessed. Thus a rural electrification scheme may be economically justified, even though the electricity has to be supplied at a loss by the utility. This normally requires a substantial financial subsidy to be made to the REB by the government. The true cost of supplying electricity to rural areas may often be more than $1.00/kWh, whereas the price to consumers may only be $0.08/kWh or less.

Sometimes the full value of the subsidies involved is hidden in the accounts of the electricity supply utility, especially in countries where there is no separate REB. This situation can arise when rural electrification costs are not distinguished from overall costs and a uniform tariff is applied throughout the country, for both urban and rural areas.

Thus, when comparing the costs of alternative techniques for rural electrification, whether grid extension, isolated diesels or photovoltaics, it is important to ensure that similar assumptions are made regarding the value of subsidies that are made available.

If it is decided to proceed with a rural electrification scheme based on PV systems, it is vital to ensure that the users of the new technology are supported by adequate arrangements for technical assistance and the supply of spare parts. This will call for good information programmes to help users understand how to get the best out of their systems and to identify faults. Technicians will need to be trained to install systems, instruct users in operation and routine maintenance, and carry out more extensive maintenance operations when necessary.

Chapter 6 — References

6.1 M.R. Staff, 'Photovoltaic Prospects for Rural Electrification', *Proc. of INTERSOL 85 World Solar Energy Conference*, Montreal 1985.

6.2 M.R. Staff, 'Rural Electrification — Solar versus Grid Extensions — Updating the Economics', *Proc. of UK-ISES Conference, Solar Energy for Developing Countries — Power for Villages*, Reading, May 1986.

6.3 S. Riphat, 'Solar Photovoltaic Development in Indonesia', paper presented at Regional Expert Seminar on Solar Photovoltaic Technology, 10-14 June 1985.

6.4 *Solar Village Indonesia*, brochure published by TUV Rheinland Institute for Energy Technology, W. Germany, and BPPT Badan Pengkajian Dan Penerapan Teknologi, Indonesia.

6.5 'Photovoltaic Rural Electrification in French Guyana', paper by Seri Renault Ingénierie presented at European Community Pilot Projects Contractors' Meeting, Nimes, France, 13/14 April 1983.

7. OTHER APPLICATIONS

7.1. Agricultural applications

Photovoltaic systems can be used for a number of applications related to agriculture. Water pumping for irrigation is the major use, as discussed in Chapter 3 of this report. Other applications include agricultural product processing (eg. grain milling), milking machinery, cattle fencing, refrigerated storage of perishables and ice production for fish preservation.

All these systems are similar in that they employ a PV array to charge a battery which then supplies the end use, either direct for DC applications or through an inverter for AC applications. Because of the high cost of the PV array and batteries, it is necessary to optimize the design of the system as a whole. It is not normally advisable to use standard AC or DC appliances, since these, although relatively inexpensive, often have poor efficiency.

Most PV manufacturers offer standard systems for battery charging, but specialized advice is generally needed for the selection of appropriate appliances, such as DC motors for grain mills, or inverters for AC systems. Several manufacturers offer PV-powered cattle fencing systems, consisting of a small PV module for charging a battery through a regulator and a standard high voltage pulse generator. These systems are beginning to find wide acceptance in areas where it is difficult to arrange for cattle fence batteries to be recharged regularly.

There are not many examples of dedicated systems for agricultural product processing, as usually the relatively small electric motors are powered from a larger system supplying a number of end-uses. A PV-powered grain mill has been operating successfully for 6 years in Tangaye, Burkina Faso. It forms part of a 3.6 kWp PV system which also includes a water pump. The grain mill has been modified twice since the time of original installation in 1979 to obtain the required degree of fineness and consistency, but the overall availability has been reported as over 90 per cent.

Although no installations for milking machines are known in developing countries, a large PV generator for a milking parlour has been built as a research project in Ireland. The 65 kWp grid-connected system built on Fota Island in 1982 provides electricity for milking machines and milk processing equipment on a large dairy farm. The performance is being closely monitored by the University of Cork and the reliability is reported to be very high (Ref. 7.1).

Another research project in Europe is the PV-powered refrigerated cold store for agricultural produce built on Giglio Island, Italy, which started operation in 1984. A 45 kWp PV array provides power for an ozonizer (for water disinfection) and a cold store of about 275 m^3. A particular feature of the compressor used in the refrigeration plant is the control system, which selects the number of cylinders to be loaded to match the power available from the PV array (Ref. 7.2).

These large PV systems for agricultural applications are still at the development phase. The smaller systems, for powering grain mills, cattle fencing, irrigation pumps, and so on, are much simpler in concept and are technically developed. They are most likely to be cost-effective in rural areas of developing countries where the following conditions obtain:

(a) There is no electricity supply from the grid and the costs and practical difficulties of running diesel engines are high
(b) The solar input is reasonably good throughout the year, with an average of at least 4.5 kWh/m^2 per day
(c) There is a need for the application for the major part of the year, as the high capital cost requires a high utilisation factor to achieve low unit costs.

Two important factors affecting the success of any attempt to introduce new technologies into the rural sectors are (i) the institutional arrangements for technical training and support and (ii) the local management of the system. It is significant that the success of the PV grain mill in Tangayè is largely attributed to the way existing practices were adapted to form a co-operative to manage the installation. This gave a sense of communal ownership, encouraging interest and concern for the success of the project.

7.2 Water treatment systems

Although not yet commercially available, several PV manufacturers are developing complete PV-powered water treatment systems. In some systems, the water is first filtered and then given a prophylactic chlorine dose, usually in the form of sodium hypochlorite, before being transferred to storage and thence to consumers. In view of the difficulties in ensuring the supply of hypochlorite and its correct use when available, some system designers are concentrating on chemical-free water treatment processes, which involve slow sand filtration for sterilization. This approach removes all harmful bacteria, but care has to be taken to avoid subsequent contamination.

In these two approaches, the PV power is used only for pumping the water, first from the source, such as a borehole or river, and then through the various filtration stages to the final storage reservoir. Another type of water teatment system being developed incorporates a PV-powered UV light for sterilization.

After the necessary development and testing has been completed, these complete water treatment packages using PV power are likely to be of particular interest for water supply projects in remote areas where the existing water sources are known to be heavily polluted. Two British-designed systems are known to be operating in Nigeria.

7.3 Telecommunications

Telecommunication systems in developing countries have traditionally been powered by grid electricity or stand-alone diesel generators. Battery banks are usually provided for security of supply in the event of power interruptions. Problems of unreliable supply, variable quality (voltage spikes, low voltage) and high cost of operation and maintenance cause constant problems for the system operators. The quality of communications frequently suffers as a result.

Photovoltaic generators are particularly suited for telecommunication systems, since they can provide the relatively small amounts of power required at remote transmission/reception sites reliably and with little or no maintenance. PV generators are widely used for this application and many hundreds of systems are operating worldwide, including in places where the average solar input is as little as 2.5 kWh/m^2 per day. In fact, photovoltaic systems have had more commercial success for telecommunications applications than any other remote power application.

Figure 7.1 PV System powering a UHF network in South America (Source: BP Solar)

There are three main types of telecommunication systems which can be PV-powered:

(i) Two-way radios, including radio-telephones
(ii) Radio and television secondary (infill) transmitters
(iii) Telephone systems, including exchanges, repeater stations and satellite ground stations (see Figure 7.1).

In addition television sets can have a dedicated PV power supply and educational TV is an application which has found wide use in certain West African countries.

In each case, the PV system is primarily required for battery charging. For the larger systems, a stand-by diesel generator may be provided, with controls for automatic start-up if the battery voltage falls to a pre-set low level. The hybrid arrangement is optimized for the least cost configuration.

PV systems are likely to be found cost-effective for sites where grid electricity is not available for loads up to about 2.5 kWh/day for most locations. In some circumstances, PV/diesel hybrid systems may be found cost effective for loads up to 10 kWh/day.

7.4 Cathodic protection

Another application where PV systems have been found substantial commercial markets is for cathodic protection of steel pipelines and other steel structures. Cathodic protection by the impressed current method involves maintaining the steel structure at a negative potential with respect to the surrounding soil or atmosphere. PV systems are particularly well suited for this application, since they provide the necessary DC power without the need for transmission lines, transformers and rectifiers as required for grid-powered systems. An example is shown in Figure 7.2.

7.5 Unusual applications

Photovoltaic systems should be considered wherever there is a requirement of small amount of power in remote or inaccessible locations. For example, aircraft warning lights on tall structures or on hilltops, or navigation lights marking out the channel into a harbour.

The eleven large steel lattice towers carrying a 230 kV double circuit transmission line 15 km long across the Jumuna River in Bangladesh are each fitted

Figure 7.2 PV-powered Cathodic Protection System for Pipeline in Abu Dhabi

with five aircraft warning lights powered by a 700 Wp PV array (Ref 7.3). Battery storage is sufficient for 10 days supply. Other methods of supplying the lights considered but rejected were diesel generators (access problems for fuel supplies), induction from the phase conductors (no power during times of outage), earth wires from the local network (unreliable).

There are now many navigation buoys and lighthouses supplied by PV generators. The cost savings can be substantial, as servicing the batteries in navigation buoys of the diesel generators associated with lighthouses is expensive and sometimes hazardous. There are five lighthouse systems ranging from 2.6 kWp to 18.2 kWp operating or under construction in the Mediterranean (Greece, Italy, France) plus several hundred PV-powered navigation lights for harbour entrances and buoys along the southern coast of France. There are similar developments in many other countries.

PV systems can also provide power conveniently and economically for remote metering installations, such as river gauging stations, meteorological stations and groundwater-level monitoring systems. The records can either be transmitted by radio to the control centre or stored on tape or disk for later collection. Some PV installations in France use satellites (Argos or Meteosat systems) for data transfer.

Chapter 7 — References

7.1 S. McCarthy, G.T. Wrixon and A. Kovach, 'Data Monitoring of the Photovoltaic Project', Presented at the first Working Session of the European Working Group on Photovoltaic Plant Monitoring, Ispra, Italy, 11-13 November 1985.

7.2 'Water Disinfection System and Cold Store', Design Report presented by Pragma at Contractors Meeting of the Commission of the European Communities Photovoltaic Pilot Projects, Nimes, France 13/14 April 1983.

7.3 D.A. Hughes and A.B. Wood, 'Jamuna River 230 kV Crossing — Bangladesh: II. Design of Transmission Line', *Proc. Instn. Civil Engineers, Part 1*, vol 76, pp 951-964, Nov 1984.

8. CONCLUSIONS

8.1 Summary of experience

General

Several thousand photovoltaic systems have been installed in developing countries over the past ten years, the great majority over the past five years. The size of these systems ranges from a few watts to over 30 kWp, for applications as diverse as water pumping, vaccine refrigeration, domestic lighting, cattle fencing and telecommunications. Many of the systems have been of an experimental nature, for developing and demonstrating the technology, but increasingly photovoltaic systems are being installed for sound commercial reasons, as being the most cost-effective solution for particular applications.

The technical, economic, social and institutional factors associated with each application need to be carefully considered before any general conclusion can be made regarding the viability of photovoltaics. Obtaining reliable data on the performance of PV systems is not easy, as there have been only a limited number of published reports on actual field experience. It is also not easy to evaluate the economic prospects, as many systems have not been operating long enough to yield sufficient data on component lifetimes and replacement costs. Social and institutional factors have a major influence on the success or failure of projects to introduce new technologies into the rural sector and many valuable lessons have been learned in the course of implementing PV projects in different countries.

Water pumping systems

About 2000 PV-powered water pumping systems have been installed worldwide, in various configurations. Most of these are for water supplies for villages and livestock watering, but some have been installed primarily for irrigation. The performance of many systems has been disappointing due to a number of factors, including:

— Use of unreliable and/or inefficient sub-system components (motors, pumps and power conditioning equipment)

— Poor overall system design, resulting in poor matching between the components in relation to the solar input and water level

— Use of inaccurate data regarding solar input and water resource conditions at the design stage.

There is evidence that the manufacturers have learned from past experience and that the latest types of pumping systems are considerably more reliable and efficient than earlier models.

On the basis of reasonable assumptions regarding the life of motor/pump units, the unit water cost from PV pumps is found to be competitive with the unit water cost from diesel pumping systems in remote areas for applications where the volume-head products (daily volume required in m^3 times the total pumping head in m) is less than about 1000 m^4. Wind pumps would probably be cheaper than PV pumps if the mean wind speed in the periods of maximum water demand is at least 2.5 m/s. Even though the unit water cost for PV pumps may be cheaper than for diesel pumps in certain circumstances, this does not necessarily imply that PV pumps would be an economic solution for irrigation applications, since the economic market price of the crop has to be considered. Nevertheless, PV pumps combined with trickle irrigation or other low-water-use irrigation techniques, when used for fruit and other high value crops, may be found to be economic. Finance in the form of capital grants and low-cost loans will be needed to bring high capital cost/low running cost equipment within reach of the potential users.

PV water pumps have found wide social acceptance, particularly in villages which previously had to pump water by hand. The full involvement of the end-users from the planning stage onwards has been found to be the key to successful implementation of PV pumping systems. In some cases, a village co-operative has been formed to administer the local aspects, such as fund raising to meet local expenses and the levying of water charges on users. A central organization, usually government-backed, is essential however to support the users with finance and technical advice. Training of the users is needed to enable them to obtain the best results and to adapt local practices as appropriate. They also need to be trained in routine maintenance and trouble-shooting. Simple faults can often be repaired by unskilled operators, with the central support organization providing help for more serious problems and obtaining spare parts.

Vaccine refrigerators

As a result of the co-ordinated development and field testing programmes sponsored by WHO, PV vaccine refrigerators are now available which perform reliably and are cost-effective for use in areas without grid electricity. The key technical factor is the sizing of the PV array and the battery, which requires knowledge of the climatic conditions and the loads likely to be experienced by the system at the proposed site.

Detailed technical requirements for PV refrigerator/freezers intended for vaccine storage as part of the 'cold chain' and other recommendations for procurement are set out in guidelines issued by the WHO. At present, nine commercial systems have been approved for use in the 'cold chain' and several other systems are being evaluated.

The average annual cost of a PV refrigerator is comparable to that for a kerosene refrigerator. PV refrigerators are more cost-effective, however, since they are more reliable, which results in a lower cost per dose. Since the refrigerator cost per dose is relatively small compared with the immunization programme overhead cost per dose, the use of the more efficient refrigerator is well justified.

Provided the PV refrigerator is properly designed, installed and maintained, it should give a reliability of at least 90 per cent (ie. only 10 per cent of vaccines lost), compared with reliabilities as low as 50 per cent for kerosene refrigerators. Immunization programmes can be seriously disrupted by vaccine losses, with courses of immunization incomplete, medical staff frustrated and loss of momentum of the programme as a whole.

The users of PV vaccine refrigerators need to be trained in the correct operation and maintenance of the system. As for PV water pumps, there needs to be a central organization available for users to obtain technical support, advice and spare parts.

Lighting

Small PV systems for domestic lighting are widely available and several thousands have been installed, particularly in the South Pacific, French Polynesia and China. They are simple to operate and reliable, now that earlier problems with battery charge regulators have been solved. PV-powered fluorescent tubes provide a much higher quality of light compared with candles or kerosene wick or pressure lamps. The efficiency of the DC ballasts used for fluorescent tube lamps is a key technical factor in the design of systems.

Larger PV lighting systems are also available for street lighting and security lighting. Standard systems using fluorescent tubes are available from most PV manufacturers. Again the key technical factor has been the efficiency of the DC ballasts. Specially designed AC systems for high-mast lighting have also been demonstrated. The key factor here is the performance of the inverter.

Based on analysis over five years with 10 per cent interest rate, PV lighting systems are found to be cost-competitive with kerosene lamps in areas where the cost of kerosene is $0.75/litre or more. Thus, given suitable finance, users would be able to save money and improve the quality of domestic lighting by installing PV lighting. Several organizations are operating successful financing schemes which provide a capital grant plus a loan for the balance repaid over five years at 9 or 10 per cent interest.

It is essential that in addition to technical advice and the provision of finance, users of PV lights have access to spare parts, particularly replacement fluorescent tubes and batteries. Some form of local co-operative, backed-up by a central organization, is likely to be the best approach in most cases.

Rural electrification

Rural electrification differs from lighting in that it is a more systematic approach to providing an electricity supply to a village or district. The electricity is then available for a wide range of domestic, commercial and agro-industrial applications. A number of relatively large central PV demonstration plants have been built to electrify a complete village, but there are a number of technical and institutional problems with this approach. Central systems of this type are vulnerable to failure due to component faults or over-loading and thus need a high level of supervision.

A more viable approach to rural electrification is to equip each household with its own stand-alone system. A number of standard sizes would be required to suit the size of the household and the nature of the loads likely to be imposed. Larger systems would be needed for commercial, institutional and industrial premises.

The electricity produced for general applications, such as lighting, fans, refrigerators, radios and televisions, needs to be AC, at the national voltage and frequency. This means that the PV systems have to incorporate inverters, which have not proved particularly efficient or reliable in the past. However, several manufacturers are now introducing new inverter types which offer much improved performance. At least one type switches itself into stand-by mode when no loads are connected, thereby greatly reducing the internal system losses. Initial field experience with these inverters is encouraging.

PV systems for rural electrification (as opposed to simple lighting applications using DC output) are too expensive at current costs for large-scale implementation. In certain circumstances, where the cost and other difficulties involved with diesel generators are too great, PV systems can be cheaper, but users would not in general be able to afford to pay the true cost, even if long-term finance were available. Substantial subsidies, comparable to the subsidies already provided for rural electrification schemes by grid extension or diesel generation, will help bring costs within the reach of consumers. However, PV system costs need to come down to about 50 per cent of present levels for this technology to be an acceptable alternative to conventional techniques.

Any scheme for rural electrification using PV systems will need to be planned to provide for public information and other forms of institutional support. This would probably be best organized within the existing rural electrification board, to cover all technical, administrative and financial aspects.

Agricultural applications

Stand-alone PV systems can be used for a number of agricultural applications which require mechanical power. Besides water pumping for irrigation, grain milling is probably the most important agricultural use. Although not many PV-powered grain milling systems have been built, there is no reason to doubt the technical feasibility of this application. One system has been operating in Burkina Faso with high reliability for some six years. The type of mill has to be selected to suit the type of grain and the degree of fineness required. The mill may then be driven by a DC motor supplied by a battery charged by a PV array. Care must be taken to ensure that all components are correctly matched to suit the solar input as it varies throughout the year, and the load imposed. Such applications are likely to be economic in comparison with diesel-powered systems in remote areas, provided the demand is reasonably uniform throughout the year.

Other PV applications for agriculture include milking machinery, refrigerated storage for perishable produce and ice production. Experience with these applications is too limited for any general conclusions to be made. One other application has however found wide acceptability technically and economically, and that is PV-powered cattle fencing. These electric fences can be used not only to keep domestic livestock within required limits, but also to keep wild animals out of areas where they could damage crops.

Water treatment

In addition to PV pumps for water supplies, PV-powered complete water treatment plants are being developed for use in remote areas. The emphasis is on reliability and minimum requirements for chemicals and other consumables. Several approaches have been demonstrated, including systems involving slow sand filtration, filtration followed by a prophylactic dose of sodium hypochlorite and filtration followed by UV sterilization. Experience is too limited for any general conclusions to be made regarding the best approach. Costs are likely to be competitive with any alternatives involving diesel generators.

Telecommunications

PV systems for telecommunications are finding wide acceptance throughout the industrialized and developing world, since they offer a reliable and cost-effective means of providing relatively small amounts of power in remote locations. Essentially, the PV array charges a battery which supplies conventional telecommunications equipment. Hybrid systems, involving back-up diesel generators, are often found to be the most cost-effective solution for larger systems (ie. daily demand greater than about 2.5 kWh).

There are three types of telecommunication systems which can be PV powered:

— Two-way radios and radio-telephones
— Radio and television secondary (infill) transmitters
— Telephone systems, including exchanges, repeater stations and satellite ground stations.

In addition, television sets can have a dedicated PV power supply, an application which has found wide use in certain West African countries for use in village schools.

Cathodic protection
PV generators for cathodic protection systems used for pipelines and other steel structures are commercially competitive in areas remote from the electricity grid. There are many examples of such systems used for oil and gas pipelines in the Middle East. This is an application where the DC output from the PV array can be directly used.

Other applications
PV systems can be used wherever there is a requirement for small amounts of power in a remote location. One manufacturer in Europe has recently developed a mobile orthopaedic clinic powered by a PV generator to enable a full range of equipment to be used in areas which have no electricity supply. Hazard warning lights on tall structures, navigation lights at harbour entrances and lighthouses are some examples of suitable applications for photovoltaics. Remote metering stations, such as river gauging stations and meteorological recording stations, can also be PV powered economically. New applications are constantly being found as the potential of the new technology becomes more widely appreciated.

8.2 The technology

In general, the photovoltaic modules and arrays have performed reliably with very few reports of failure or significant degradation. The experience to date has been solely with crystalline silicon solar cells, both mono-crystalline and semi-crystalline. This is because most manufacturers manufacture modules which meet the qualification and performance requirements laid down by the Jet Propulsion Laboratory (USA) or the CEC's Joint Research Centre (Italy). The performance of the newer types of thin film amorphous silicon solar cells, which are beginning to become available for power applications, has yet to be evaluated.

Many early PV systems (ie. prior to about 1982) suffered problems with power conditioning and control systems, such as voltage regulators and maximum power point trackers. This was due to a number of factors, including inadequate weather protection, over-complex design and not being sufficiently robust, both physically and electrically. There is growing evidence that the power conditioning and control equipment now being supplied for use with PV systems is much more reliable, but further development to improve the efficiency of some devices, particularly inverters, is required.

For systems that incorporate batteries, it is vital that the correct type of battery is chosen and that the sizing calculation takes full account of actual operating conditions. Automotive batteries are in general not suitable, as they have limited life when subjected to many charge/recharge cycles. A number of battery manufacturers have developed special batteries for PV applications which offer long life and low internal losses and require little or no maintenance.

The end-use devices, such as motor/pump units and refrigerators, have generally had to be specially developed for PV applications. Many problems developed with the earlier systems, often due not so much to faults in the overall concept, but to mistakes in the matching of components, faulty operation and inadequate quality control. New products with better performance, reliability and durability have become available in recent years for all applications of main interest in developing countries.

There has been extensive work to test and evaluate PV pumping systems and PV vaccine refrigerators. There is a continuing need to update the results of previous projects and to keep potential customers informed of the state-of-the-art. Similar testing and evaluation programmes are needed for other PV systems, such as lighting systems, electrification systems suitable for households and institutional buildings (health centres, police posts, schools, etc), and systems for agricultural product processing such as grain mills.

8.3 The economics

It is not possible to generalize about the economic viability of PV systems. Each application has to be considered on its merits, taking into account local conditions and the cost of alternatives. Although PV systems have high initial cost, they require no fuel and little maintenance and should last many years. In many remote areas, diesel generators, the main alternative to PV generators, would be impracticable due to fuel supply costs and uncertainties, plus the problems associated with maintenance and the supply of spare parts. Even if diesel generators appear to be cheaper on a life-cycle cost comparison, it might be preferable to go for a PV system because of the operational advantages.

For some applications, PV systems are widely found to be competitive with the alternatives. For example, PV refrigerators for vaccines offer a lower cost per dose than kerosene refrigerators and enable a more effective immunization programme to be mounted. PV systems for domestic lighting are also competitive with kerosene lamps and candles and moreover give a much better light.

PV water pumps may be cost-competitive with diesel pumps for applications where the flow is low (less than 50 m^3/day) and the head is low to medium (less than 20m). The analysis is sensitive to the cost of diesel fuel and the life-expectancy of the diesel pump. PV pumps are more likely to be viable for village water supply or livestock water supply applications, where the social benefits are high, than for irrigation applications, where the extra value of the crop made possible by the irrigation water has to exceed the cost of the pump.

PV systems for telecommunications and cathodic protection are often found to be the cheapest option when small amounts of power (up to about 10 kWh day) are required at sites remote from public electricity supplies. Depending on local circumstances, PV systems may also be the cheapest alternative for hazard warning lights, remote metering stations, navigation buoys and lighthouses.

8.4 Social and institutional factors

Experience has shown that PV systems are generally widely welcomed by the users, provided they have been involved at an early stage in the planning process and have been given basic instruction in how to operate the system and carry out routine maintenance. Problems arise when a system is set down in the field by a research organization and the local people are expected to use it without any real appreciation of what is involved and without proper arrangements for follow-up.

Unlike diesel systems or grid-supplied electricity, PV systems are very site-and load-dependent. Therefore considerable experience and technical expertise are needed at the design and procurement stage in order to ensure the system will fulfil its expectations. A central organization within the country concerned needs to be established with the necessary skills in-house to undertake the necessary design tasks and write system specifications in preparation for tendering procedures. The same central organization can arrange for user training and technical support in the operation phase. The supply of spare parts, particularly where these have to be imported, is an important function of this central organization.

There are four important institutional factors that contribute to the success of a PV application:

(a) Genuine involvement by the users from the planning stage onwards, with appropriate arrangement for a co-operative or other method of administering the local aspects (operation, maintenance, fund raising, user charges, etc)
(b) Technical expertise at the system design and procurement stage to ensure that the system is compatible with local needs, solar resources and other technical constraints
(c) Finance to meet the majority of the initial costs, after allowing for any government subsidies considered appropriate, with repayments spread over at least five years at a preferential interest rate
(d) Arrangements for user training and technical support after installation, to deal with faults that cannot be fixed locally and the supply of spare parts.

Although most developing countries have already established a department to undertake the necessary supporting activities as outlined above, many need technical assistance and finance from outside to help them identify and implement appropriate photovoltaic projects.

9. RECOMMENDATIONS

9.1 Identification of appropriate applications

Photovoltaic applications need to be evaluated taking into account all relevant technical, economic, social and institutional factors. The results of this evaulation then form the basis for subsequent decision making. The technical evaluation will need to include consideration of the following:

- operating performance of the system as a whole in relation to demand and solar resource variations with time;
- operating performance of each main component with a view to identifying possible improvements;
- reliability, availability and durability;
- ease of operation, maintenance and repair.

The economic evaluation will need to take into account:

- initial costs for procurement, transportation to site and installation, including any local civil engineering costs, subdivided into foreign exchange and local currency costs;
- labour costs for operation and maintenance;
- expected costs of spare parts and replacement of components which wear out;
- life cycle cost-benefit economic analysis, based on appropriate discount rate;
- unit cost comparisons with alternative energy sources;
- financial analysis from user's viewpoint, taking into account subsidies,loans, interest rates and repayment periods.

The evaluation of social and institutional factors will need to take into account the following:

- the actual demand for the energy and/or product from the system;
- the social and environmental context in which the system will operate;
- the level of operator skills available and associated training needs;
- the appropriate form of local organization needed to administer the system;
- the appropriate form of central organization needed to provide technical and financial support.

Only when satisfactory answers to each part of the evaluation are obtained should the project proceed to the next stage in the decision-making process.

9.2 Strategic approach to development

A strategic approach to development is essential if the limited resources of finance and technical skills are to be utilised in the most effective ways. There are many competing demands on national resources and national policies have to be developed which balance these demands in the manner considered most appropriate in the national interest.

Photovoltaic systems can contribute to development in a number of sectors. Village water supplies and lighting systems can significantly improve the standard of living of the rural population, thereby raising morale and helping to counteract the drift of young people to the towns. Irrigation pumps can increase agricultural output, thereby providing more food for home consumption or for export, as well as raising rural incomes. More reliable vaccine refrigerators can improve the effectiveness of immunization programmes and thereby reduce infant mortality. Improved and extended telecommunication systems can contribute significantly to raising efficiency in all sectors of national life.

PV projects must be considered alongside other development projects and ranked in order of the net benefits likely to accrue to the nation as a whole, taking into account both economic and social benefits. The effects the project would have on matters such as employment, food production, balance of payments and development of national skills and self-reliance need to be assessed. In this connection, the possibilities for local manufacture and assembly of as much of the PV system as practicable need to be considered.

These issues are not easy and straightforward to evaluate. They involve many factors that are hard to quantify, but it is important to appreciate that the viability of PV projects should not be assessed simply in terms of economics or major energy substitution. For example, the energy requirements for telecommunications are relatively small but the benefits that can result from installing reliable PV generators in place of unreliable diesel generators can be of very great significance.

9.3 Staged development

Solar energy activities, in common with all work in the field of appropriate technology, need to be planned with the ultimate objective of introducing suitable systems into the community on a commercial basis, either with or without government subsidies. The full development sequence should therefore be planned along the following lines:

Stage 1 — Research to identify the basic physical principles, materials and designs; most of this activity in relation to photovoltaics has been (and will continue to be) carried out in the industrialized countries, as it involves high technology equipment for making and testing solar cells and other components;

Stage 2 — Laboratory-based development work to adapt system designs to suit local materials and needs, to characterize performance under local conditions and to identify how performance can be improved; such work is very useful for training professional and technical personnel in the principles of solar engineering and test methods;

Stage 3 — Field work on pilot plants, to demonstrate and test representative systems under realistic field conditions, preferably when used by local people such as villagers or farmers; at this stage, it is important to monitor the systems in detail and to involve potential industrial companies who may be interested in subsequent commercial development;

Stage 4 — Full-scale demonstration plants, with prototype commercial units installed at a number of representative sites throughout the country, with the full involvement of the industrial interests and continued technical support from the research institution responsible for the original development.

Stage 5 — Commercialization, with local manufacture/assembly of systems and associated marketing and follow-up activities. In addition to the commercial suppliers of hardware, a separate independent organization is essential for several years to promote schemes and to provide technical support to the users.

Usually financial support from the government is required for stages 1 and 2. Costs for stages 3 and 4 are often shared between the government and commercial interests, with possibly some contribution from the users. For Stage 5, government subsidies and finance for low-interest loans are often needed to bring high initial cost systems within the reach of potential users in rural areas.

Taking into account the experience now available from many countries, the photovoltaic applications of main interest in developing countries are at the following stages of development:

— water pumps	Stages 4 and 5
— vaccine refrigerators	Stages 3, 4 and 5
— lighting systems	Stages 4 and 5
— rural electrification (central plants):	Stages 3 and 4
— rural electrification (household systems):	Stages 4 and 5
— agricultural systems	
— grain mills	Stage 3
— cattle fences	Stage 5
— milking machines	Stage 3
— cold stores	Stage 3
— telecommmunications:	Stage 5
— cathodic protection:	Stage 5
— hazard warning lights:	Stage 5
— lighthouses:	Stage 4

Each developing country needs to assess its own needs and institute a staged development for each PV application of interest, taking into account the status of development reached in similar situations elsewhere. An essential requirement is to build up the necessary institutional support with the skills and finance needed to implement the programme.

www.ingramcontent.com/pod-product-compliance
Lightning Source LLC
Jackson TN
JSHW060309140125
77033JS00021B/629

* 9 7 8 0 9 4 6 6 8 8 3 9 5 *